MECHANICAL TYPEWRITERS

Their History, Value, and Legacy

Thomas A. Russo

Schiffer Publishing Ltd
4880 Lower Valley Road, Atglen, PA 19310 USA

Library of Congress Cataloging-in-Publication Data

Russo, Thomas A., 1932-
Mechanical typewriters: their history, value, and legacy/by Thomas A. Russo.
p. cm.
ISBN 0-7643-1545-5
1. Typewriters—History. 2. Typewriters—Collectors and collecting. 3. E. Remington & Sons—History. 4. Remington Rand, Inc.—History. I. Title.
Z49.A1 R87 2002
652.3'09—dc21

2001006598

Designed by John P. Cheek
Cover design by Bruce M. Waters
Type set in Korinna BT/Korinna BT

ISBN: 0-7643-1545-5
Printed in China
1 2 3 4

Published by Schiffer Publishing Ltd.
4880 Lower Valley Road
Atglen, PA 19310
Phone: (610) 593-1777; Fax: (610) 593-2002
E-mail: Schifferbk@aol.com
Please visit our web site catalog at **www.schifferbooks.com**
We are always looking for people to write books on new and related subjects. If you have an idea for a book please contact us at the above address.

This book may be purchased from the publisher.
Include $3.95 for shipping.
Please try your bookstore first.
You may write for a free catalog.

In Europe, Schiffer books are distributed by
Bushwood Books
6 Marksbury Ave.
Kew Gardens
Surrey TW9 4JF England
Phone: 44 (0) 20 8392-8585;
Fax: 44 (0) 20 8392-9876
E-mail: Bushwd@aol.com
Free postage in the U.K., Europe; air mail at cost.

CONTENTS

ACKNOWLEDGMENTS

Upon returning from a tour of duty in the U. S. Navy during the Korean War, I immediately went to work for Remington Rand in Kansas City, Missouri. My job was selling typewriters. I was very fortunate to work directly for a person that would have a profound impact on the remainder of my life. William C. Matthews was my sales manager and he was responsible for teaching me how to be a strong competitor in the work place, while at the same time exercising a high degree of business ethics and fair play.

Matthews eventually became worldwide Director of Marketing for Remington Rand, at that time a division of Sperry Rand. Matthews was not only a positive inspiration to me, but to just about everyone he came into contact with at Sperry Rand.

On April 1, 1971, I left Remington Rand to form my own company in Wilmington, Delaware as a dealer for Remington products. I was unaware that on that same day, Matthews was planning to do the same thing, but in Memphis, Tennessee.

Bill and I became involved in the trade association of our industry (NOMDA). Working together, we accomplished much to aid our dealers and manufacturers. At the same time, we built successful and respected companies in our local areas. Bill's two sons, Bill, Jr. and Mark, are presently guiding their company, "Business Equipment Center of Memphis," through the digital age. Once again our lives have paralleled each other. My two sons, Thomas James and John Christopher, have acquired control of my Delaware company, now called "D.O.E. Technologies, Inc." I am pleased to say that my grandson, Thomas Robert Russo, is now working in the business with his father and uncle.

Remington Rand and the industry it helped spawn have been very good to Bill and our families as well as me. So, I am dedicating this book to that industry, that company and the wonderful people who worked side by side with us. But especially to William C. Matthews, Sr., who will always be "Mr. Typewriter" to me and thousands of others who had the honor of knowing and working with him.

I would also like to thank another former employee of Remington Rand, Mr. David P. Sheridan. David was the Worldwide Advertising Manager for Remington and has made available a great deal of material on the subject of the Remington Company and its products.

Once again, I would like to thank Mary Jo, my wife; and immediate family (Thomas; Kathleen; Callye; Mary Ann; Cheryl and John, my children) for their help and encouragement through all three books on the subject of early Office Technologies. An example of their support was demonstrated by my granddaughter, Anna Russo, who used her artistic abilities to create a typewriter quilt for my July 26, 2001 birthday.

Anna's Quilt – July 26, 2001.

I would also like to thank Mrs. Susan Belock and Mrs. Nina Wolf for their assistance in editing the text of this book.

And finally, a special thanks to my longtime friend and associate, Sonya Zawalski, who has helped me type, format, and proof not only this book but the two previous books as well.

INTRODUCTION

With the merger of three dynamic forces, a great worldwide typewriter industry developed. First was the invention of a practical typewriter product that was faster, easier, and better than all previous forms of writing. Few will argue with the fact that the Sholes and Glidden Typewriter with its universal "QWERTY" keyboard was that product. It was not the first typewriter but it was unquestionably the typewriter that changed the way people and organizations would communicate for the next one hundred plus years. The second force was a powerful international company with the manufacturing and engineering know how that could transform the primitive Sholes & Glidden typewriter into a marketable product; a company with worldwide distribution channels and the financial strength that was needed to guide this product through its perilous beginning. That company was a fifty-eight year old successful gun manufacturer, E. Remington & Sons. Finally, the third force, and probably the most dynamic of all, was women. Previous to the introduction of the typewriter, women were primarily involved in such occupations as schoolteachers, dress designers, seamstresses, and factory workers. Women were well-suited to the typewriter and skillfully mastered its operation. They would eventually increase speed and efficiency to the point unimagined by typewriter designers.

This book will attempt to track these three dynamic forces and the wonderful office revolution they inspired. There were hundreds of typewriter manufacturers and thousands of products that surfaced during this period. The author will share some of these national treasures with the reader, as well as attempt to provide dates of manufacture, patent numbers, and inventor information, where available.

Also included are estimates as to availability and value. The value is based, in some cases, on the price paid by the author for these artifacts and also on prices realized at auctions, private sales, and other antique marketplaces. The reader should be aware, however, that while price is based in some cases on availability, it is always based on condition. The cost of restoring an old typewriter by a qualified specialist can often dwarf the cost of the machine. Cost is finally based on how badly the seller wants to sell and how badly the buyer wants to buy.

Chapter One
EARLY BEGINNINGS

The early beginnings of the typewriter could conceivably go back as far as the beginning of civilization when man began to communicate and draw pictures in the sand; or, perhaps to the writings of the Babylonians, the Egyptians, or even the Romans. However, the establishment of an alphabet was the foundation upon which future progress would be based. It formed the foundation for writing and making a permanent record of how we speak or communicate in our native languages.

At this point, perhaps unconsciously, man began his search to write as fast as he could speak. Obviously, this was a challenge not to be achieved for thousands of years and would require many innovations.

Paper

Sir Aurel Stein discovered the first traces of paper at the Great Wall of China. There, in a spur of the Wall, he found what was described as state documents that dated between 21 – 137 A.D. It was therefore obvious that the Chinese had developed paper to a degree that was nearly perfect at this early date. The Chinese taught the art of papermaking to the Arabs sometime in the 8th century. They in turn introduced the art to Europe in the 13th century.

Ink

In the 4th century, using lamp-black, the Chinese discovered ink. Red ink made of sulfide of mercury was used as far back as the Han Dynasty.

After its discovery, red ink was used exclusively by the Imperial Family or Royalty and black ink was used by the common people. Black ink was also found to be especially adaptable for use on wooden blocks (see *Reverse Imaging* below).

Reverse Imaging

The discovery of reverse imaging dates back to the first use of seals or marks when traders made an impression in clay to show the items origin. Around the 5th century, these seals were moistened with ink before the impression was made. In the 7th century, Buddhist missionaries experimented with various methods of duplication utilizing seals, stencils, and textile prints.

So, ink, paper, and seals led to the development of type and printing. Printed blocks were discovered in Japan that date back to 770 A.D. The world's first book, *Diamond Sutra,* was printed by Wang Chiek on May 11, 868 A.D. While the *Diamond Sutra* was printed on a scroll, the same discovery included fold books with the words "a new development" included. These books were printed with wooden blocks.

The 10th Century, saw the development of paper money printed by the Chinese. In 1297, Marco Polo described with amazement the respect the Chinese displayed for these curious scraps of paper. It was not until 1657 that Europe first issued paper money.

In 1041, also in China, Pi Sheng formed movable type made of earthenware. Since the Chinese language had 40,000 individual characters, very little use was made of this discovery.

The Koreans, in 1403, produced the first metal type known to history. The first models were engraved in hard wood and molds of porcelain paste were made from these models, which were then baked in an oven. From these, the metal type was cast.

The basic rudiments were now in place: ink, paper, and reusable type. Now the concept of mechanical transportation could revolutionize the publication of books and create an industry of printing.

While some circles claim that Laurens Janzoon Koster of Holland was the first to invent the printing press, others, with equal enthusiasm, claim that Pamfilo Costaldi of Italy was the first. But the consensus of the general world is that a German printer and inventor from Mainz, was the inventor of the printing press. Johannes Gutenberg, in 1437, developed a method of printing from movable type, which included the use of molds, metal alloys, a press, oil-based inks, and a methodology of automatically impressing on paper. A permanent record was created which could be printed and reprinted simply by re-inking. Each type was also reusable. It was not necessary to create a new mold every time a new character was printed. The first book printed with the new press was the Bible. A few of the original copies of the Gutenberg Bible remain and are still considered to be masterpieces.

In 1501, an Italian by the name of Ottaviano dei Petrucci, used movable type to print musical compositions.

While books, proclamations and musical compositions could be printed and reproduced automatically through the mechanical innovations of the period, ordinary letters or correspondence for individual use remained a task that was accomplished laboriously through the use of the quill or pen and ink. The process was slow and messy, not to mention prone to errors. Copies were not readily available. Thomas Jefferson, the American author of the "Declaration of Independence," invented a machine that used two pens side-by-side. The second pen produced a copy of the document as a by-product of the original.

Cartoon of Queen Anne.

It was not until 1714 that progress was made. An Englishman, Henry Mill, is credited with the first patent for a device described as "an artificial machine for the impressing or transcribing of letters, singly or progressively one after another as in writing." There is no surviving model or machine of the Mill invention. Only a description survives in the patent documentation.

The Italians figured prominently in the development of the printing press and the device to print musical compositions. They were equally prominent in the development of the actual typewriter. Pellegrino Turri, Pietro Conti, and Guiseppe Ravizza, all made contributions to a machine that would type letters. Probably, the most outstanding of these three Italians was Ravizza. His invention in 1837, called a "Cembalo-Scrivano," was the first of several models. His patent model won recognition at the Exhibition in Turin, Italy in 1858.

As early as 1772, John Saur had established the first successful American type foundry in Germantown, Pennsylvania, a small community outside of Philadelphia. In 1836, David Bruce, invented the first American type casting machine.

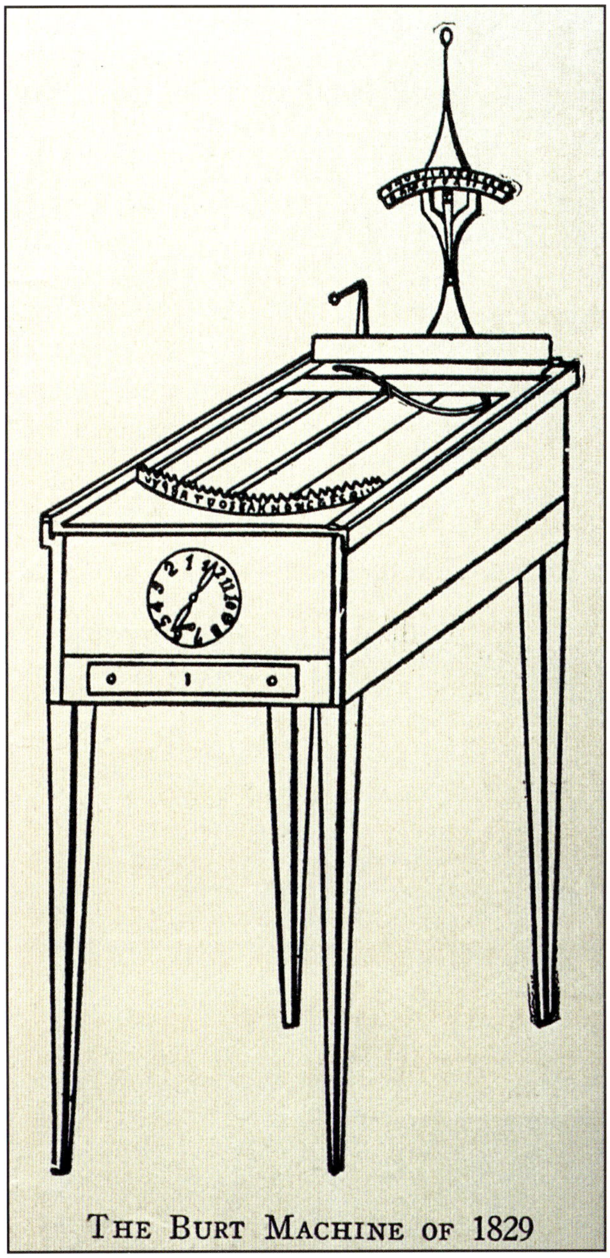

Burt's first patent.

William Austin Burt (June 13, 1792 – August 18, 1858) was the first American inventor of a typewriter-like device, which he called a "Typographer." His 1829 model looked more like a modern pinball machine than a typewriter. But nevertheless, his patent number 259, dated July 23, 1829, is recognized as the American grandfather of our typewriter inventions. Burt also invented a Solar Compass in 1836 and an Equatorial Sextant in 1856.

In 1881, Edward H. Knight published a *Mechanical Dictionary* in three volumes. In Volume III, Knight listed early typewriter inventors, their patent numbers and dates as follows:

Patent Number	Inventor	Date
3,228	Thurber	August 26, 1843
4,271	Thurber	November 18, 1845
7,052	Fairbank	September 17, 1850
7,771	Eddy	November 12, 1850
8,990	Jones	June 1,1852
10,995	Thomas	May 30, 1854
14,907	Cooper	May 20, 1856
14,919	Jones	May 20, 1856
15,164	Beach	June 24, 1856
18,504	Francis	October 27, 1857
22,428	Harger	December 28, 1858
38,815	De Mey	June 9, 1863
39,296	Livermore	July 21, 1863
57,182	Peeler	August 14, 1866
59,522	Flamm	November 6, 1866
62,206	Johnston	February 19,1867
65,807	Hall	June 18, 1867
79,265	Sholes	June 23,1868
79,808	Sholes	July 14,1868
81,000	Pratt	August 11, 1868
87,941	Johnson	March 16, 1869
94,329	Moore	August 31, 1868
109,161	Washburn	November 8, 1870
115,287	Draper	May 30, 1871
118,491	Sholes	August 29, 1871
124,437	Halstead	March 12, 1872
125,952	Hansen	April 23, 1872
127,789	Cadmus	June 11, 1872
133,841	Edison	December 10, 1872
139,914	Pember	June 17, 1873
140,921	Hill	July 16, 1873
144,450	Galloway	November 11, 1873
148,940	Gally	March 24, 1874
158,071	Hansen	December 22, 1874
168,868	Hansen	October 19, 1875
169,757	Allisoff	November 9, 1875
170,233	Case	November 23, 1875
170,289	Crandall	November 23, 1875
170,621	Deming	November 31, 1875
171,139	Johnson	December 14, 1875
171,835	Allen	December 31, 1876
171,408	Morgan	December 21, 1875

Also see the following English Patents:

Number	Date
395	1714
9,204	1841
9,745	1843
10,939	1845
868	1854
1,110	1854
674	1856
1,050	1856
1,868	1857
101	1863
2,031	1863
295	1864
893	1864
3,983	1864
2,260	1864
3,257	1864
1,412	1864
2,687	1864
1,186	1866
1,448	1866
8,163	1866
866	1869
997	1869
3,699	1869
3,324	1870
918	1871
3,177	1871
823	1872
990	1872
1,912	1872
2,161	1872
2,239	1872
2,481	1872
3,548	1872
668	1873
3,852	1873
927	1874
1,786	1874
2,181	1874
3,528	1874
2,119	1875

French Patent:

Number	Date
7,028	1840

While all of these inventions were conscious attempts to advance the technology and are therefore important, a few are described in more detail.

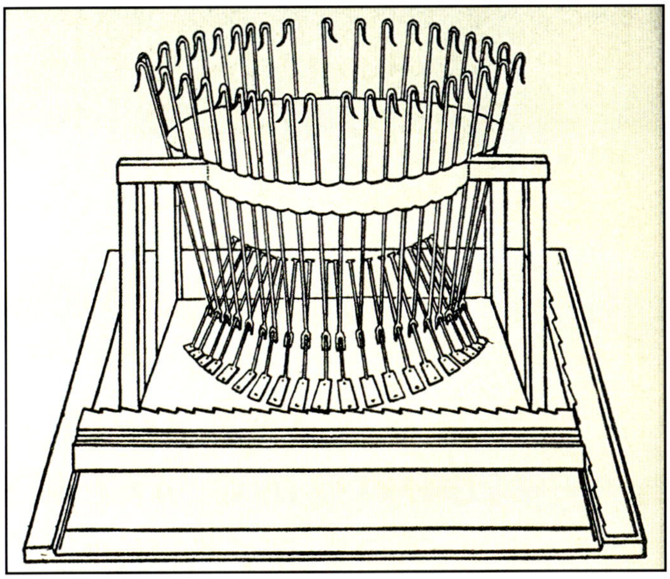

Progin Patent.

In 1833, Xavier X. Progin, an inventor from Marseilles, France, made substantial progress with his device called a "Typographic Machine or Pen." Many of the mechanical principles found in Progin's machine were incorporated in the advancement of the typewriter technology. His major contribution was a circle of "type attached to bars" that were attached to hooked wires that caused the type to strike on a flat bed of paper. The entire instrument moved from left to right and when a line was complete, it then moved down to the next line.

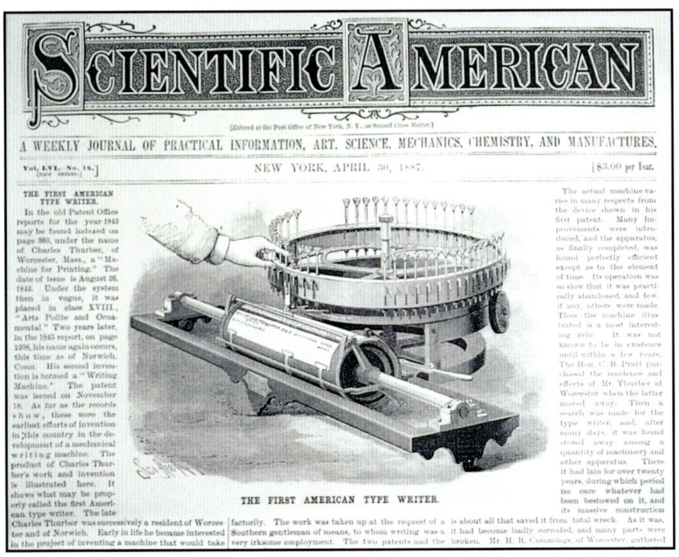

SCIENTIFIC AMERICAN

A WEEKLY JOURNAL OF PRACTICAL INFORMATION, ART, SCIENCE, MECHANICS, CHEMISTRY, AND MANUFACTURES.

Vol. LVI.—No. 18. [NEW SERIES.] NEW YORK, APRIL 30, 1887. [$3.00 per Year.

THE FIRST AMERICAN TYPE WRITER.

In the old Patent Office reports for the year 1843 may be found indexed on page 360, under the name of Charles Thurber, of Worcester, Mass., a "Machine for Printing." The date of issue is August 26, 1843. Under the system then in vogue, it was placed in class XVIII., "Arts Polite and Ornamental." Two years later, in the 1845 report, on page 1208, his name again occurs, this time as of Norwich, Conn. His second invention is termed a "Writing Machine." The patent was issued on November 18. As far as the records show, these were the earliest efforts of invention in this country in the development of a mechanical writing machine. The product of Charles Thurber's work and invention is illustrated here. It shows what may be properly called the first American type writer. The late Charles Thurber was successively a resident of Worcester and of Norwich. Early in life he became interested in the project of inventing a machine that would take [...] factorily. The work was taken up at the request of a Southern gentleman of means, to whom writing was a very irksome employment. The two patents and the [...]

The actual machine varies in many respects from the device shown in his first patent. Many improvements were introduced, and the apparatus, as finally completed, was found perfectly efficient except as to the element of time. Its operation was so slow that it was practically abandoned, and few, if any, others were made. Thus the machine illustrated is a most interesting relic. It was not known to be in existence until within a few years. The Hon. C. H. Pratt purchased the residence and effects of Mr. Thurber at Worcester when the latter moved away. Then a search was made for the type writer, and, after many days, it was found stored away among a quantity of machinery and other apparatus. There it had lain for over twenty years, during which period no care whatever had been bestowed on it, and its massive construction is about all that saved it from total wreck. As it was, it had become badly corroded, and many parts were broken. Mr. H. R. Cummings, of Worcester, gathered

THE FIRST AMERICAN TYPE WRITER.

Thurber's Machine – *Scientific American* – April 30, 1887.

Ten years later, in 1843, another American, Charles Thurber (January 2, 1803 – November 7, 1886), designed a typewriter that incorporated type arranged in a circle and attached to the ends of a rod with a key top at one end and the type at the other. By moving the type in a circular manner and stopping at a universal point, the depression of the key top made an impression on a cylindrical platen that moved back and forth and by revolving caused a vertical movement. This "platen and carriage" innovation would also be used on future typewriters.

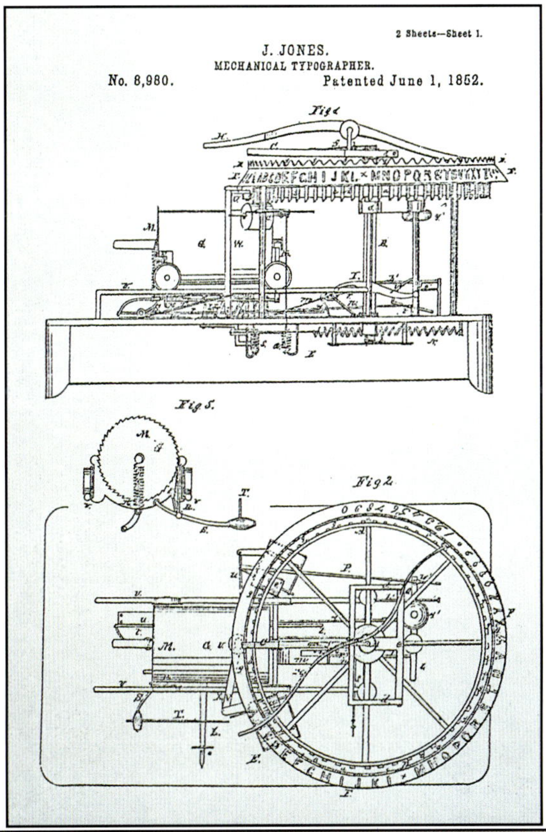

Fifth typewriter invention in the U. S. was this machine patented by John Jones in 1852. It incorporated the first practical use of vertically set characters. The toy typewriters made for children today still follow the principle of the Jones machine.

On June 1, 1852, John Jones was granted a patent number 8,980 for a machine that was beginning to resemble what we know today as a typewriter. It was called a "Mechanical Typographer." It was also one of the first typewriters to be manufactured.

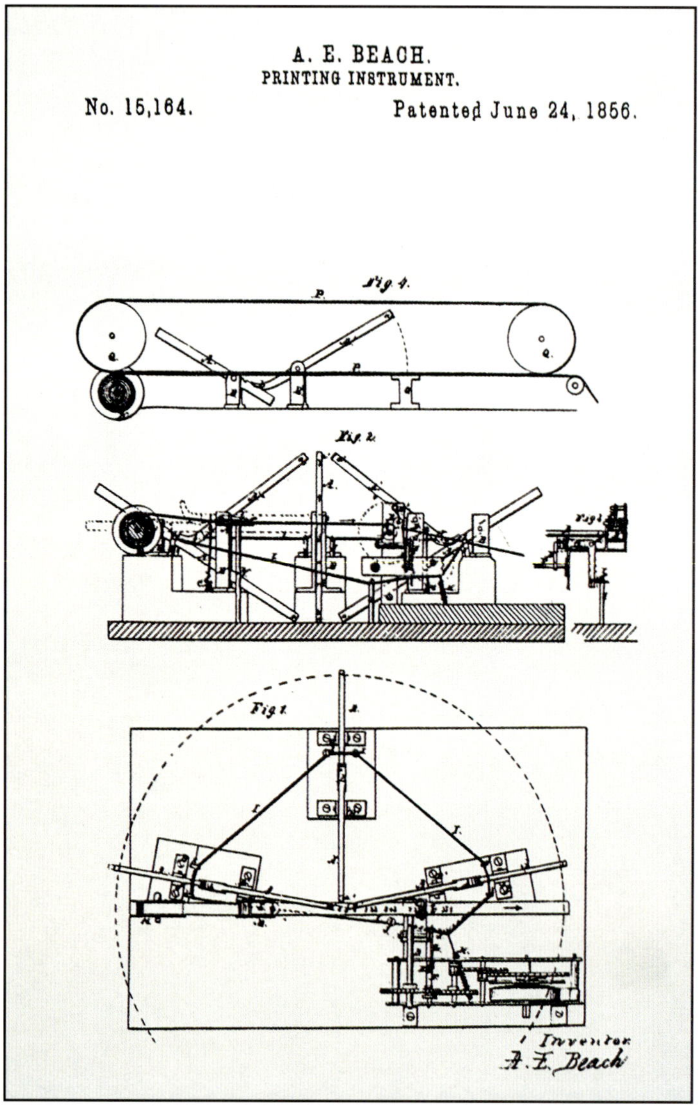

Progress now began to move more rapidly. In 1856, Alfred E. Beach (September 1, 1826 – January 1, 1896), the editor of the *Scientific American* invented several machines. One of which is pictured as it appeared in an international trade paper and was reproduced in the 1923 edition of "Typewriter Topics." Also shown is Beach's patent number 15,164 dated June 24, 1856. Beach was part of a partnership that owned and operated America's foremost scientific magazine, the *Scientific American*. In addition to the above patent, Beach held patents for a typewriter for the blind, cable railways, and pneumatic tubes for mail. He used the same principle to project a train-like device carrying passengers through a tunnel.

Some typewriter inventions had keyboards that resembled a piano more than they did a typewriter. The Wheatstone (c.1851) and the Francis (c.1857) were examples using such a keyboard.

Left & below: Alfred E. Beach machine and his patent number 15,164 dated June 24, 1856.

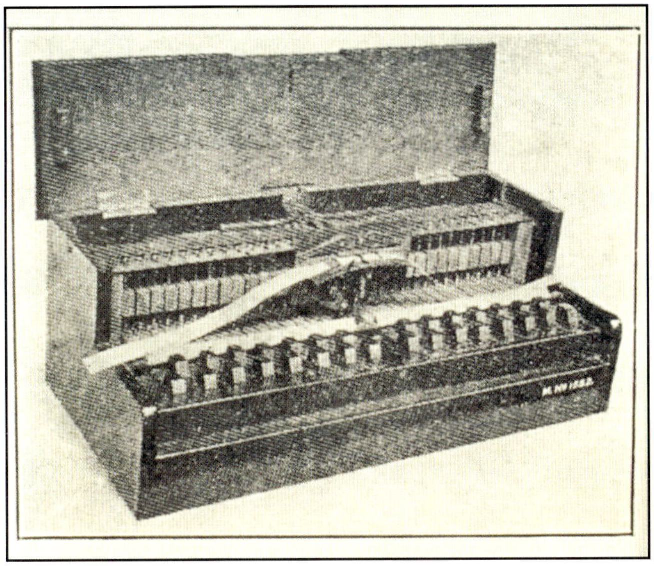

Wheatstone typewriter of 1851.

Francis typewriter of 1857.

English inventor and scientist, Sir Charles Wheatstone (February, 1802 – October 19, 1875) was born in Gloucester. As an entrepreneur at the age of twenty-one, he began his own business manufacturing musical instruments. He did research work and published papers on the subjects of acoustics, light, and optics. The "stereoscope" was one of Wheatstone's ideas that entertained families throughout the world. He developed a polar clock that made it possible to tell the time of the day by light from the sky even though the sun was hidden from view.

He was best known for his work with the electric telegraph. Although he was not the inventor of the electric telegraph, he was the person most responsible for making it available for the public transmission of messages. He was the first to gauge the velocity or speed of electricity passing through a copper wire. These experiments led to many new inventions, such as a letter-showing dial telegraph in 1840, the type-printing telegraph in 1841, and the magnetoelectric dial telegraph.

Wheatstone had an uncanny ability to translate hieroglyphics and decipher coded messages. As a result of this dexterity, he invented a cryptograph writer. For many years his code was unbreakable. For these and many other contributions to science and invention, Wheatstone was knighted on January 30, 1868.

A physician and author who was exceptionally gifted in mechanical innovation was Samuel Ward Francis (December 26, 1835 – March 25, 1886). Coming from a long line of physicians, Francis graduated from medical school on October 27, 1859. In the same year, he was awarded a patent for a "Printing Machine" device that was designed to be a typewriter. "The principal feature of the patent was that he arranged a row of hammers in a circle, so that when put in motion they would strike in the same place. Then he connected a piano-like keyboard to the device. The paper was moved along by means of a spring and catch so connected with the keys that it should move the paper a distance of one letter whenever a key is struck. At the end of each line the 'car' which carries the paper is drawn back by the hand." While many of the features of the Francis machine were complicated and sluggish, the one outstanding feature that would be used on later machines was the "type guide." During his life, Francis was awarded sixteen patents, which included a sewing machine (1875), a signal for telephone and telegraph lines (1879), and a device for canceling postage stamps (1863). In 1876, the U. S. Government announced they had saved $200,000, the annual salaries for 254 clerks, by using a stamp-canceling device.

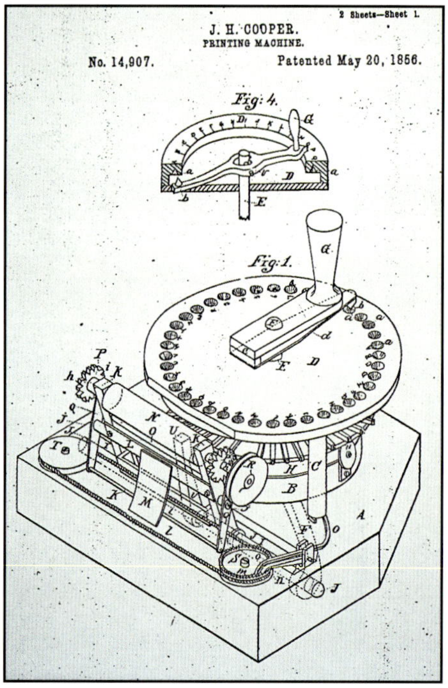

J. H. Cooper Machine. Patent number: 14,907 dated May 20, 1856.

Pratt's machine of 1868. Patent number: 81,000.

In 1856, John H. Cooper of Philadelphia, Pennsylvania, developed an index typewriter that made a stamping impression by selecting a letter or character and then hitting rather than depressing the lever.

John Pratt of Centre, Alabama, was a journalist by trade. During a period in which he resided in England, he became interested in developing a typewriter. Because he had little mechanical experience, he hired an English model maker to help develop his ideas. He completed his invention in 1864. On December 1, 1866, he was granted patent number 3,163 in London, England, for a typewriter that he called a "Ptereotype."

John Pratt (April 14, 1931 – 1905).

The Pratt typewriter was important in many ways. First of all, it was the best effort to-date. But even more importantly, it was a description of Pratt's invention in the *Scientific American* magazine that inspired or provoked another American to develop what would become the world's most popular typewriter.

Christopher Latham Sholes.

The before mentioned article in the *Scientific American* proved to be exceptionally interesting to a former Milwaukee, Wisconsin, printer and editor. At the time he was an employee of the Federal Government as a collector of customs at the Port of Milwaukee. That man was Christopher Latham Sholes, the "Father of the American Typewriter."

Kleinsteuber's Machine Shop located in Milwaukee, Wisconsin.

Kleinsteuber's was a favorite meeting place for amateur inventors. Sholes often went there for help in building models of his ideas. Sholes and a close friend, Samuel W. Soule, had previously patented a machine for addressing newspapers. A few years later, Sholes patented a machine to number pages.

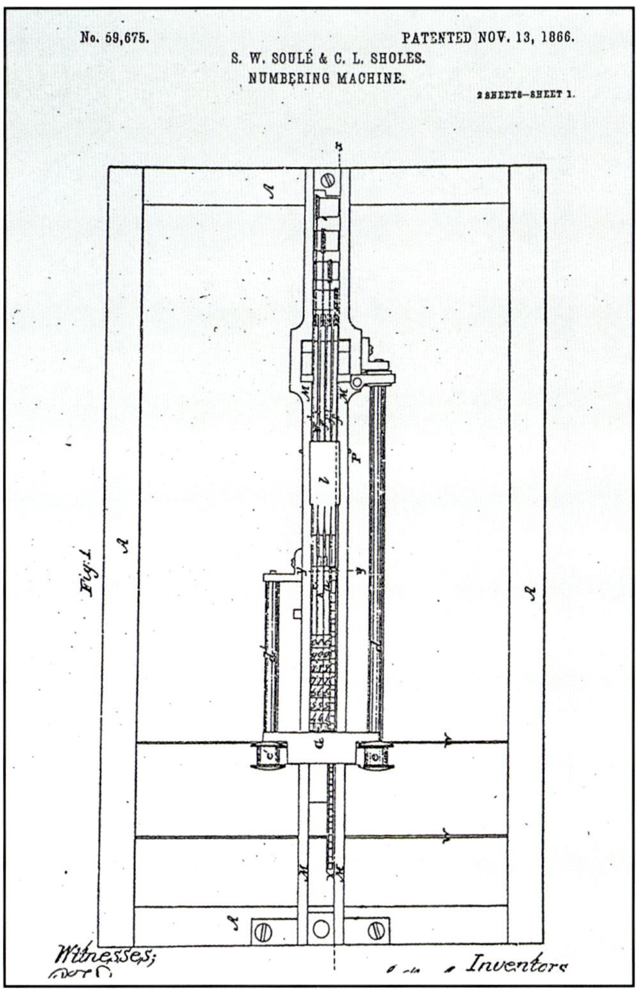

S. W. Soule and C. L. Sholes patent number 59,654 dated November 13, 1866 for an addressing machine.

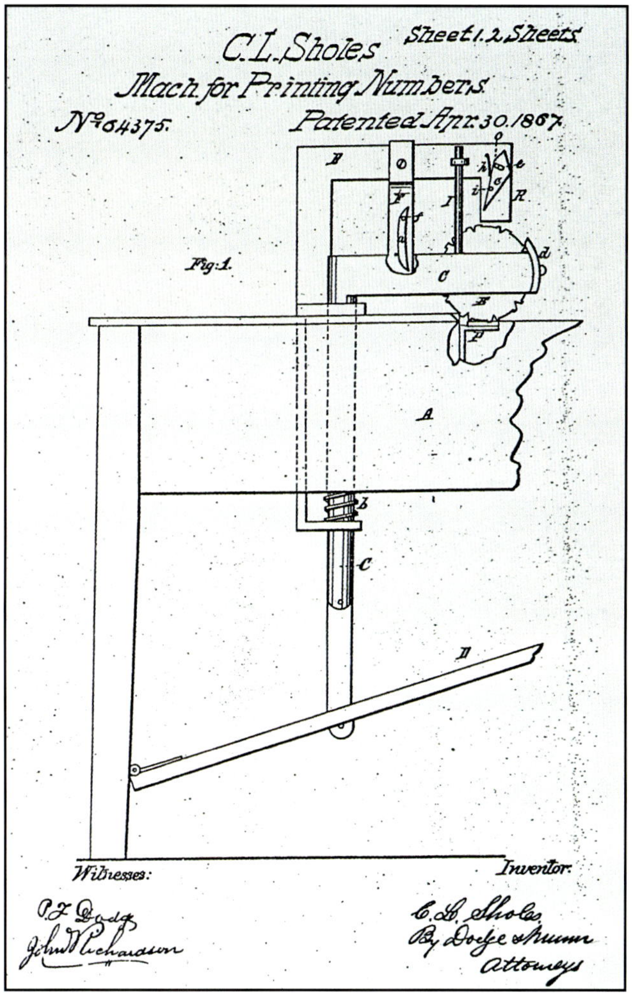

C. L. Sholes patent number 64,375 dated April 30, 1867 for a numbering machine.

After seeing the numbering machine, another friend of Sholes, Carlos Glidden, suggested that if Sholes could build a machine to type numbers, why not a machine to print letters. That's exactly what he set out to do. Initially, during the development period, he had help from Carlos Glidden, Samuel W. Soule, and Matthias Schwalbach.

Sholes was the one who contributed most of the innovative ideas and by the fall of 1867, their first model was completed. The first patents granted to Sholes are as follows:

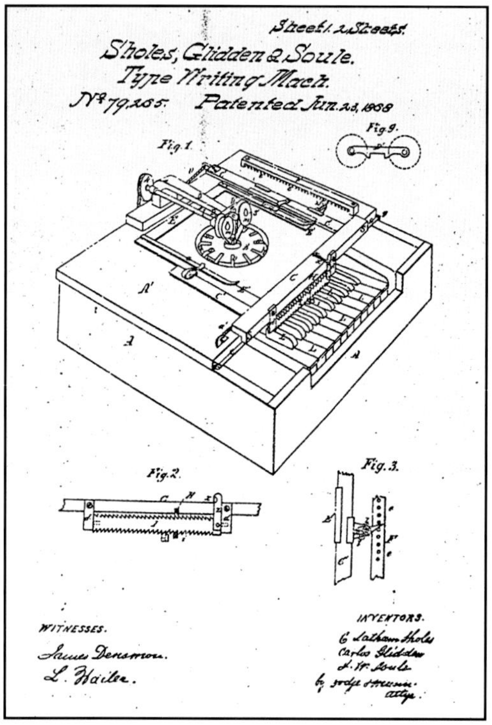

Sholes Patent number: 79,265 – applied for – June 23, 1868.

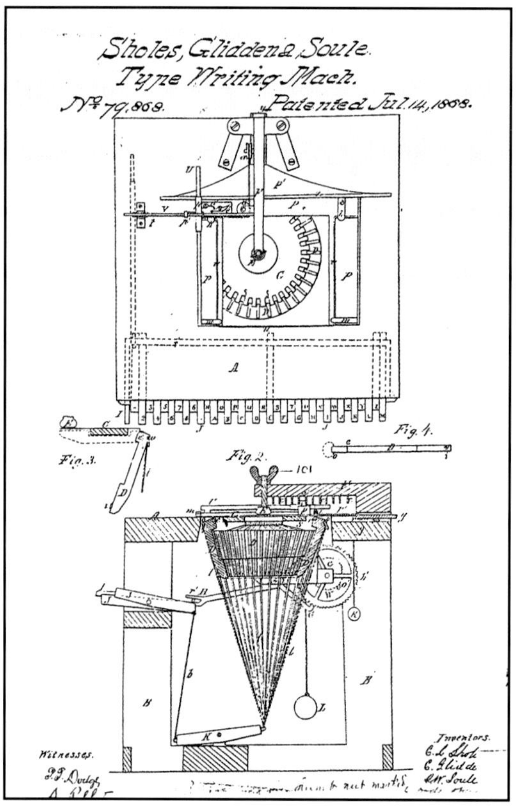

Sholes Patent number: 79,868 – applied for – July 14, 1868.

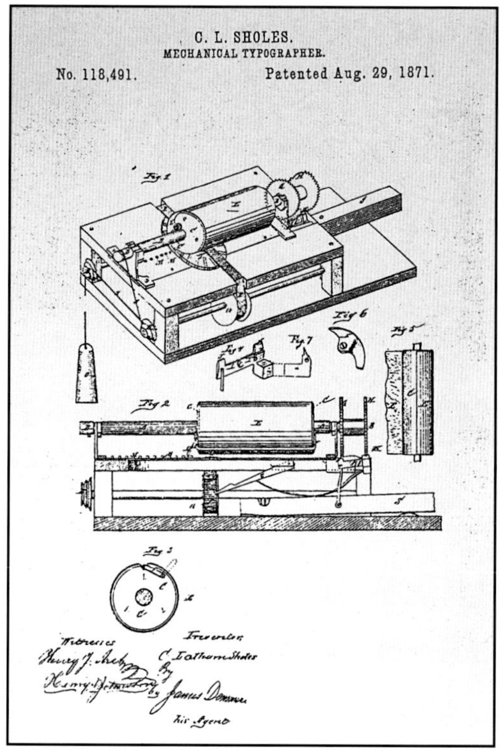

Sholes Patent number: 118,491 – applied for - August 29, 1871. The Sholes Axle Model. Developed in 1868 and patented two years later.

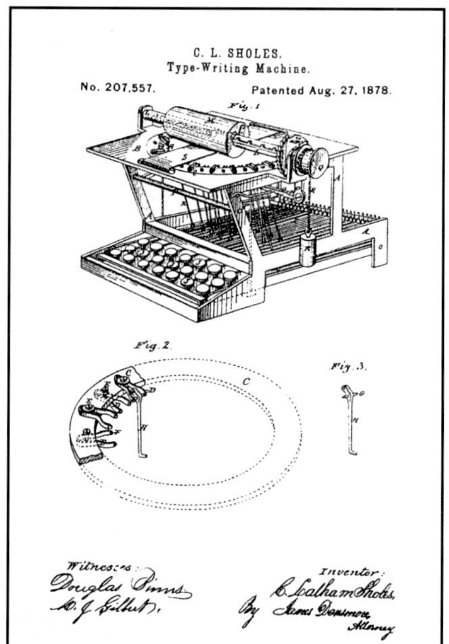

Sholes Patent number: 207,557 – applied for - August 27, 1878. The Sholes Axle Model fully developed and manufactured by Densmore in 1871.

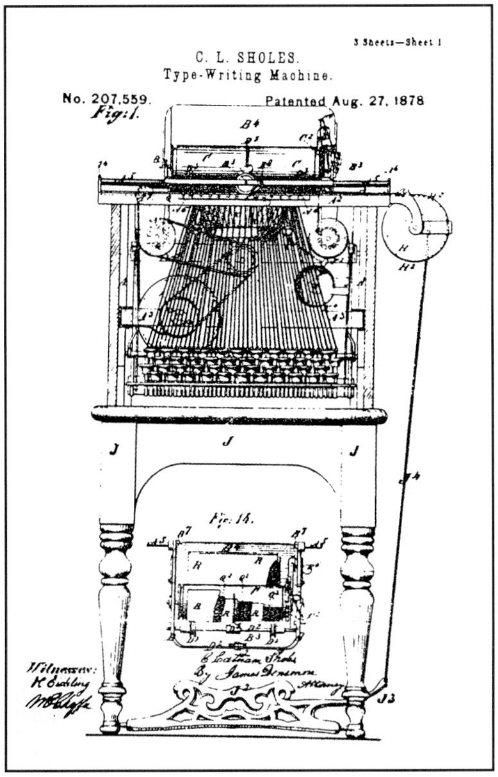

Sholes Patent number: 207,559 – applied for - August 27, 1878. Machine of 1872 patented six months later.

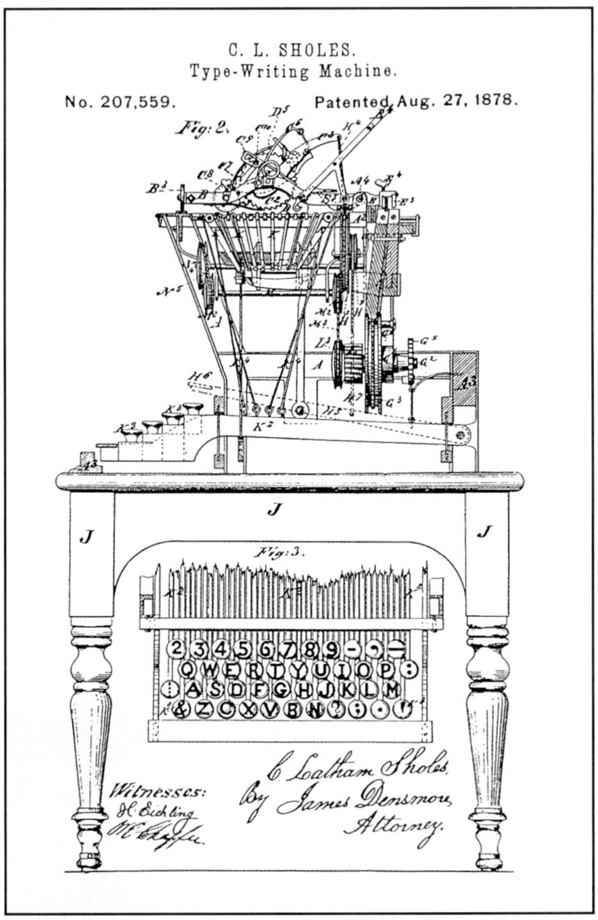

QWERTY keyboard. (See figure three) Patent number: 207,559, dated August 27, 1878.

Sholes worked tirelessly in improving his invention. There were problems with the ribbon inking system, problems in making carbon copies, and problems with jamming type bars at the point of print. The latter problem was resolved not only by trial and error but also by suggestions from James Densmore. The final arrangement of the keyboard appears in the patent below. From then on it would be known and accepted universally as the QWERTY keyboard. It became so ingrained in the minds of operators that in spite of many efforts to replace it with keyboard arrangements that were more scientific and faster, the QWERTY keyboard is still with us today on our modern computer keyboards.

By 1872, Sholes felt the typewriter was ready to be manufactured and released to the public. Densmore agreed with Sholes. After demonstrating the machine to a friend, George Washington Napoleon Yost, the question of how best to manufacturer and market this new invention was discussed. Yost immediately suggested a gun manufacturer located down in the Mohawk Valley along the banks of the Erie Canal. Densmore quickly wrote a letter to Philo Remington, President of E. Remington & Sons. Densmore briefly described their new typing device and requested a meeting to demonstrate its virtues and discuss a potential mutually profitable relationship.

The results of that meeting would be the beginning of the Remington Typewriter Company as well as the start of an industry that would eventually employ millions of men and women throughout the world.

The Sholes invention marked the end of the early typewriter pioneers and the beginning of a new Industry. There were many wonderful typewriter inventors after Christopher Latham Sholes, who contributed substantially to the industry. But, the early typewriter inventors up to and including Sholes, deserve the honor and distinction that has been given to them. Their struggles are reflected in their marvelous patents and their efforts will never be forgotten.

The Roman poet and philosopher, Titus Lucretius Carus (c.99 B.C. – 55 B.C.) in the fifth volume of his six volume set *De rerun natura* (The Nature of Things) expressed their efforts worthily and it bears repeating here. His comments were also included in the frontispiece of Knight's *American Mechanical Dictionary*. It is an appropriate and fitting tribute to the efforts of the early typewriter inventors and pioneers:

> "Thus time brings all things, one by one, to sight,
> and skill evolves them into perfect light."

Chapter Two
AN AMERICAN MANUFACTURER APPEARS: E. REMINGTON & SONS

One of the three major forces responsible for the success of the typewriter was, strange as it may seem, a manufacturer of firearms. The story of E. Remington & Sons and the events that led to the development of its first rifle could go back as far as the beginning of civilization.

Early examples of weapons. Cave man with a slingshot and hunter with a crossbow.

Mans first attempt at propelling a missile or bullet through the air started with man throwing a rock at an animal or adversary of the time. He soon realized that the distance that he could throw a rock was limited by the strength of his throwing arm. So he developed the first primitive weapon, the slingshot, a device with which he became amazingly accurate. Years later, David, the Jewish shepherd boy who became the King of Israel, slew the giant Goliath and demonstrated this accuracy.

It is hard to believe that thousands of years later in the year 2001, with all the technology available, one can turn on a television set and witness young Palestinian boys using this same device against the Israelis, perhaps because it required little manufacturing resources, a patch of leather, a piece of twine, and a small rock. The material was and is still readily available.

The need to do more than bruise an adversary at a greater and safer distance produced another weapon also easy to make. The necessary components were again very accessible and led to the invention of the bow and arrow, then long bow, and, finally, the crossbow. However, these were still examples of primitive weaponry.

The science of weaponry was introduced with the discovery of gunpowder. There is some evidence that the Chinese were the first to discover gunpowder. However, in the thirteenth century, a monk named Roger Bacon, recorded in his notes that, "*11.2 parts of saltpeter mixed with 29.4 parts of charcoal and 29 parts of sulphur*" produces an explosion.

Over the next several centuries, heavy cannons, inaccurate guns and related clumsy weaponry slowly advanced. Finally, at the beginning of the 17th century, the flintlock appeared and ammunition in the form of a cartridge was invented.

The cartridge opened the door for the breech-loader. The American Civil War began with muzzle-loaders and ended with breech-loaders. This brings us to the Remington family in America.

Forty-six years before the Civil War began – around 1815, on a small farm located near Ilion, New York, a young boy asked his father if he would give him money for a rifle. At that time, this was not an unusual request for a boy to ask of his father. But the father said "no;" also, not an unusual answer. The father's name was Eliphalet Remington I, an ironsmith, and the boy, was Eliphalet Remington II, called "Lite" by his family. Like many boys of that period, Lite would not take "no" for an answer. Rather than ask his father again, he set about making his own rifle.

Collage showing E. Remington's portrait; his first forge; and working on boring his rifle.

After several weeks of work during which his father and several neighbors offered suggestions, Lite completed the barrel for his new rifle. It was a .45 caliber, which was normal for the time, and was 42 inches long. There was one more job required to complete the barrel and Lite had neither the tools nor the expertise to do it. The rifle barrel had to be reamed. Lite had heard of a young gunsmith in Utica, New York, about fifteen miles away, named Morgan James. So he set out on foot to Utica. There James began the task of reaming and rifling Lite's barrel. James was impressed with the remarkable job that Lite had done. As a result he took extra time to see that the reaming and rifling were equally well done. Lite watched the entire process, absorbing all the details of the gunsmith's process and skill. When the barrel was completed, Lite took it back home where he completed the hand-carved stock.

Samuel Remington.

At last, the rifle was finished and it was time to put it to the test. After several shots, Lite had sighted the rifle and was repeatedly shooting bulls eyes. Feeling pretty good about the results, Lite entered a local county shooting contest. Although he did not finish first, he did finish second against some of the best shots in the area. As a matter of fact, the man who won first prize was so impressed with Lite's new rifle, he asked if he could try it. A few shots convinced him that he had to have one. He asked Lite to make one for him. Lite agreed and that was the beginning of one of America's greatest firearms and ammunition manufacturers.

Eliphalet Remington III.

Word of mouth advertising was, and still is, of tremendous help to a beginning enterprise. The Remingtons were rewarded by their good reputation as gunsmiths. At that time, all gun barrels were imported from Europe, a costly and time-consuming process. As a result, orders for Remington guns came in from all over the state.

Philo Remington.

Eliphalet II married a neighbor, Abigail Paddock, on May 12, 1814. Over the years, Abigail and Lite would have six children (three boys and three girls). The three boys would all eventually become part of the business.

Construction of the Erie Canal began in 1817, and by 1825 it was finished. The Remington's realized that if they moved to land by the canal they could take advantage of tremendous savings in shipping costs. In 1828, the Remington's purchased 100 acres of land along the south bank of the Mohawk River and across from the canal. There the Remington factory was built and there it still stands today. It later became part of what would be called Ilion, New York. 1828 was a difficult year for the Remingtons. Eliphalet I died as a result of a carriage accident. That same year Lite and Abigail moved to a new house near the factory and canal.

The business continued to prosper. By 1839, Lite's son, Philo became a full partner of the business. The company's name was changed to E. Remington & Son. They received their first government contract in 1845, during the war with Mexico.

In 1851, Samuel and Eliphalet III joined the business. Again the name was changed, this time to E. Remington & Sons. The company was formed with a well-balanced management team that was destined to guide it into becoming a world class arms manufacturer.

In 1861, the United States suffered the worst crisis of its history, the Civil War. Remington was called on to supply a major part of the weaponry (guns, pistols, and bayonets) for the Union Army. The commitment the Remingtons made to the Federal Government would not only require all of their time and effort, but all of their financial resources as well. Plant expansion and new heavy machinery was financed in order to enable the Remington plant to produce over 1,000 rifles per day. The strain of bringing the plant up to the challenge would take a terrible toll on the health of Eliphalet II. Lite died on August 12, and his three sons carried on the work.

On April 10, 1865, the war ended and the Government cancelled all contracts, including any unfilled orders. This meant that the Remington plant would have to shut down. The citizens of Ilion found themselves unemployed. All the machinery that had been financed to produce weapons during the war still had to be paid for. The local Ilion Bank would fail if those notes went unpaid. While the end of the war was a God-sent blessing, the consequences were grave for both the Remingtons and the Ilion community.

Philo met with Samuel and Eliphalet III and they agreed to use their personal financial resources to meet all of the bank obligations. This noble gesture depleted their personal fortunes.

They still had the agriculture business operation. During 1865 and 1866, it carried the company. Farm machinery had been very profitable for them until competitors started up in the mid-west. As they began losing market share because of the high cost of shipping to the mid-west, their profits declined rapidly.

They knew that they had to diversify. They tried everything, including cotton gins, fire engines, iron bridges and other things. They financed many new inventors and lost thousands of dollars. One such invention was an improved sewing machine. Its failure cost the Remington's over a million dollars.

In their search for diversity, Philo Remington took seriously a letter he received in February of 1873 from James Densmore. Densmore's letter requested a meeting to demonstrate a new device called a "Type Writer" invented by Christopher Latham Sholes of Milwaukee, Wisconsin.

Philo Remington agreed to the meeting, completely unaware that the result would revolutionize offices, not only throughout America, but also the world. Communication between businesses and individuals would never be the same.

Picture of the typewriter taken to E. Remington by J. Densmore and G. W. N. Yost and a January 29, 1945 advertisement from Time Magazine that simulated that historic meeting.

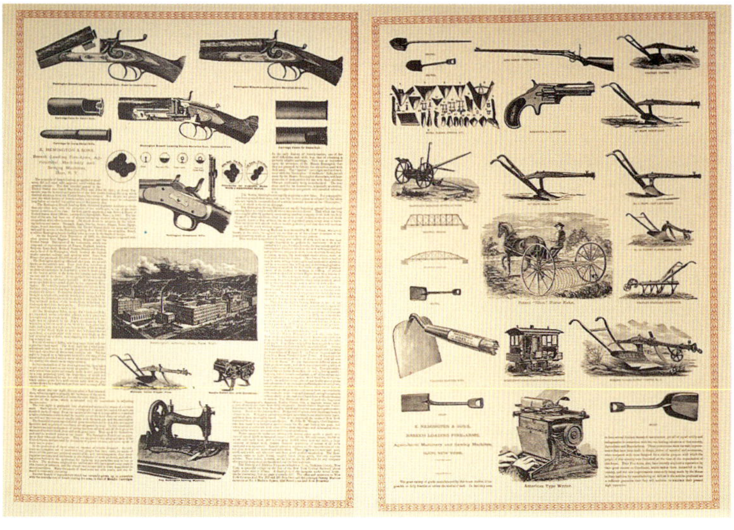

Left: History of American Industry by Archer & Descher. This advertisement depicts a post-Civil War Remington and their attempt to diversify.

The meeting took place at the Osgood Hotel in Ilion, New York. In addition to Philo Remington, attending the meeting were: Henry Harper Benedict; Superintendent of the Armory, Jefferson M. Clough; and the Assistant Superintendent in charge of the Sewing Machine Department, William K. Jenne. Representing the Sholes & Glidden typewriter was James Densmore. Densmore brought along a professional salesman, George Washington Napoleon Yost, whose background was in the oil business.

After several hours of discussion with each man trying the keyboard and typing a few words, Jenne was visibly taken by the machine and was having a difficult time hiding it. As they left the meeting, Philo Remington asked Benedict what he thought of it. His response was that Remington should not let this opportunity slip away from them. That same afternoon, Remington signed the contract to manufacturer the typewriter. As part of the negotiations, Densmore and Yost would become the first Sales Agents. Jenne and Clough were assigned the task of making the Remington more appealing from a sales point of view or, in today's terms, more user-friendly.

The Remington on a sewing machine stand.

Remington Type-Writer No. 1.

By September, just six months after the initial meeting, Jenne completed the Remington Type-Writer No. 1. It was later put on a sewing machine stand. A foot pedal was used to move the carriage back from the end of one line to the beginning of the next. It was felt that women would be among the operators of the new typewriter, so Jenne had decorative flowers stenciled around and on top of the machine. The rationale behind the decorations was that the women would be less intimidated by the fact that it was a machine, an idea that had worked well with sewing machines.

During the last three months of 1873, only eight machines were completed. They were put on the market the following year in 1874. That same year, Mark Twain, a very popular writer of the time, became one of their first customers.

In spite of the Mark Twain purchase and product endorsement, sales were very disappointing throughout 1874 and 1875. The September 1876 Centennial Exposition in Philadelphia was coming and the whole country would see this wonderful new invention called a typewriter.

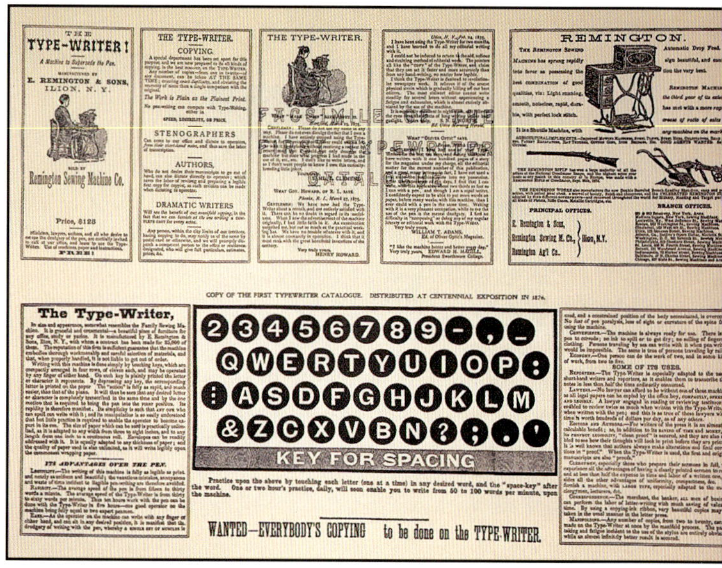

The first catalog ever prepared for the Remington Typewriter. It was distributed at the Centennial.

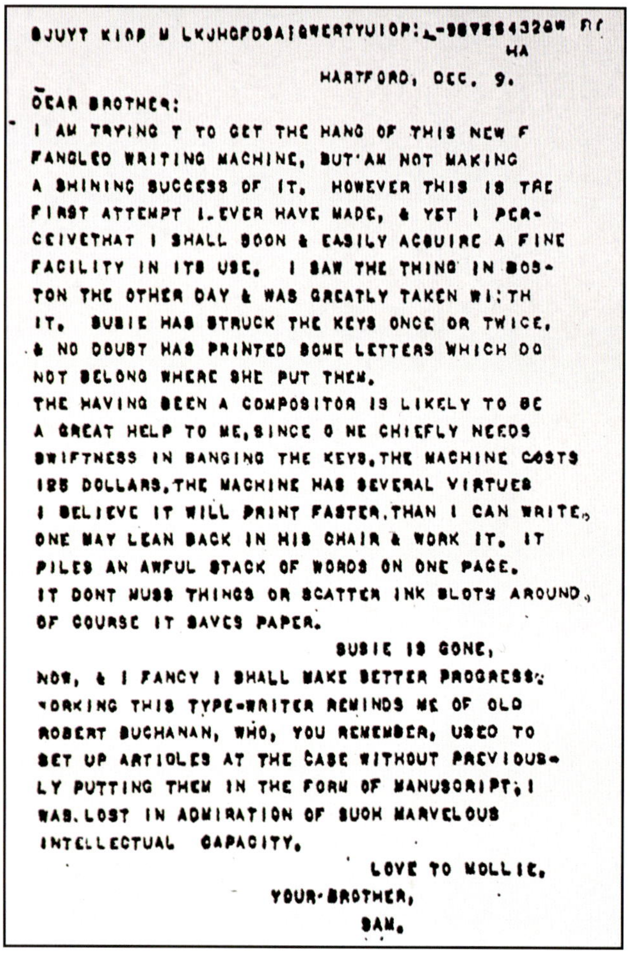

HARTFORD, DEC. 9.

DEAR BROTHER:

I AM TRYING T TO GET THE HANG OF THIS NEW F FANGLED WRITING MACHINE, BUT AM NOT MAKING A SHINING SUCCESS OF IT. HOWEVER THIS IS THE FIRST ATTEMPT I EVER HAVE MADE, & YET I PERCEIVETHAT I SHALL SOON & EASILY ACQUIRE A FINE FACILITY IN ITS USE. I SAW THE THING IN BOSTON THE OTHER DAY & WAS GREATLY TAKEN WITH IT. SUSIE HAS STRUCK THE KEYS ONCE OR TWICE, & NO DOUBT HAS PRINTED SOME LETTERS WHICH DO NOT BELONG WHERE SHE PUT THEM.

THE HAVING BEEN A COMPOSITOR IS LIKELY TO BE A GREAT HELP TO ME, SINCE O NE CHIEFLY NEEDS SWIFTNESS IN BANGING THE KEYS. THE MACHINE COSTS 125 DOLLARS. THE MACHINE HAS SEVERAL VIRTUES I BELIEVE IT WILL PRINT FASTER THAN I CAN WRITE. ONE MAY LEAN BACK IN HIS CHAIR & WORK IT. IT PILES AN AWFUL STACK OF WORDS ON ONE PAGE. IT DONT MUSS THINGS OR SCATTER INK BLOTS AROUND. OF COURSE IT SAVES PAPER.

SUSIE IS GONE, NOW, & I FANCY I SHALL MAKE BETTER PROGRESS. WORKING THIS TYPE-WRITER REMINDS ME OF OLD ROBERT BUCHANAN, WHO, YOU REMEMBER, USED TO SET UP ARTICLES AT THE CASE WITHOUT PREVIOUSLY PUTTING THEM IN THE FORM OF MANUSCRIPT. I WAS LOST IN ADMIRATION OF SUCH MARVELOUS INTELLECTUAL CAPACITY.

LOVE TO MOLLIE.

YOUR BROTHER,

SAM.

Mark Twain portrait and a copy of Twain's letter of recommendation.

Selling samples of "type-writing" for 25¢ each at the Philadelphia Centennial Exposition in 1876.

Remington booth at the Centennial.

Remington hired a model to sit at the typewriter. For a fee of 25 cents (which included postage), the model would type a short note on the new typewriter, for any customer, that they could send home.

Unfortunately, another device was released at the Centennial that overshadowed the typewriter. It was Alexander Graham Bell's telephone.

These first few years were a learning experience for Remington. They quickly found out that there was resistance to the new concept of typing. It was generally felt that the price of $125 was too high. The most important objection was that the first Remington typewriters printed only capital letters. This may have been suitable for invoicing or billing applications, but not for private correspondence.

The latter problem was corrected in 1878. Two Remington employees by the names of Lucian S. Crandall and Byron A. Brooks developed and patented a mechanical method of putting two characters on the same type lug and shifting for capital letters. This feature was available on the Remington Model 2 placed on the market in 1878. It was probably at this point that the typewriter passed the experimental stage and became practical. Remington was always very careful throughout their history to refer to their machine as the "first practical typewriter."

Remington Model 2.

In 1880, after five full years on the market, only 1,000 machines had been sold. Remington felt it could do better selling the typewriter by their direct organization. So, in 1881, that's just what it did, and Clarence Walker Seamans, one of their brightest employees, was put in charge. Seaman's recruited William O. Wyckoff, another Remington salesman and Henry Harper Benedict, an enthusiastic attendee at the original meeting presenting the typewriter to the Remingtons. The three young men soon formed the partnership "Wyckoff, Seamans and Benedict," and contracted with Remington to take all the typewriters that Remington could manufacture. That year 1,400 machines were sold. By 1882, they had sold 2,300 machines. In 1885, 5,000 machines were sold or more than sold in all previous years. At last, there was light at the end of the tunnel. Just when Wyckoff, Seamans and Benedict were at a high degree of enthusiastic optimism, a devastating rumor brought them quickly down to earth. The rumor was that E. Remington & Sons was going to sell the Typewriter Department.

Benedict's partners, Wycoff and Seamans, agreed that Benedict should waste no time in setting up a meeting with Philo Remington. Within a few days, Henry Benedict walked into Philo's office. After a brief greeting, Benedict immediately came to the point. He asked if the rumor of the eminent sale of the Typewriter Department was true. Philo confirmed that, unfortunately, the rumor was true. He then explained in detail to Benedict, the reason for the sale. Their debt service was so serious that they had no other choice.

Benedict proved to be a man of honor. He advised Philo against this course of action and explained the recent positive direction of the typewriter sales. He also said, that in his opinion, the money that Philo would receive from the sale of the Typewriter Department would not go very far in satisfying his debts. Philo indicated that there was no other option for them. They had to have the money in the next few months. It was only at this point, that Benedict told Remington that he wanted to buy the Typewriter Department.

Later, on that same day, the Remington Typewriter Department was sold to Wyckoff, Seamans, and Benedict for $186,000.

Benedict was proven correct when he told Remington that the money from the sale would not go far. It, in fact, lasted only a month. Finally, Philo asked the court to appoint receivers in bankruptcy. When the smoke cleared, Remington's liabilities were $450,000, a great deal less than their assets.

Within a few years (March, 1888), E. Remington & Sons, was sold to Marcellus Hartley of U.M.C. (a cartridge company) and Thomas G. Bennett of Winchester Arms for $200,000.

Thirteen months later, on April 4, 1889, Philo Remington died in Silver Springs, Florida.

Chapter Three
THE WOMEN'S MOVEMENT

"I feel that I have done something for the women who have always had to work so hard. This will enable them more easily to earn a living." – Christopher Latham Sholes.

Objectively reviewing the history of the typewriter, one has to ask the question, "Did Christopher Latham Sholes do something for women or did women do something for the typewriter?" Perhaps, both did something for each other. To understand the role women played in the success of the typewriter, one must understand the long struggle for women's rights and the opportunity the typewriter presented to that movement.

To understand the magnitude of the problem facing women for equal rights, Cardinal Richelieu of the seventeenth century gives us insight to the attitude that was prevalent toward women at that time when he offered the following advice to King Henry XIII: "He should give women no voice in government, for they follow their moods and passions rather than their reason. However, intellect in a woman is unbecoming; I have never seen women of much learning who was not marred by her knowledge. Women cannot keep secrets, and secrecy is the soul of statesmanship."

Therefore, it is not surprising that during the 1790s, some countries still believed that women should not be educated beyond eight years of age. Social unrest had created revolutions in France and America, and turmoil in England. Men had asserted new rights and freedoms for themselves. But women were left behind without the same freedoms and opportunities as their counterparts. In the Constitution of France in 1791, women were not considered to be citizens, as were men. Women were still thought to be frivolous and irrational. There were only a few occupations open to women. But during this same period, a woman emerged that many believe sparked the beginning of British and American feminism.

From English Roots an American Movement

Mary Wollstonecraft (1759 – 1797) was born in London to a father who was an abusive alcoholic and failed at everything he attempted. He took his frustrations out on his family, terrorizing both his wife and his children. By the time she was 19, she taught herself to read. Shunning the traditional role of women who depended on a father or husband for financial dependence, she left home and went to work.

After working for a short period, she was called back home to take care of her mother who was near death. When only 22 years old, she took complete charge of her family. When she was 25 years old, Wollstonecraft, her sister, and a friend opened a school just outside of London. After two years, the school was struggling financially and Mary began to write, initially about a subject in which she was experienced. Her first book, published in 1787, was *Thoughts on the Education of Girls*. She may have been the first woman author to publish a book.

In 1788, she published *A Vindication of the Rights of Man*. The theme was human liberty for the oppressed working class. It was during this period that Wollstonecraft went to France to observe the Revolution first hand. It resulted in another publication, *An Historical and Moral View of the Origin And Progress of the French Revolution*. In 1792, she wrote *A Vindication of the Rights of Woman*, taking previous ideas that were expressed in the rights of man and applying them to a woman.

Wollstonecraft believed that men and women were created equal, both morally and intellectually. However, she went on to say, "currently, women are inferior to men intellectually." She felt that this inferiority was due to unequal and inferior education of women. Her arguments centered on equal education for men and women and the autonomy of women. The failure of men to include women in the quest for human rights caused her to express herself in anger, outrage, and frustration that brought about a backlash of criticism.

Just five years after writing *Vindication*, William Godwin and Mary Wollstonecraft give birth to a daughter who they also named Mary. She would grow up to marry Byron Shelley and, later, would write the classic *Frankenstein*. Ten days after Mary's birth, at the age of 38, Mary Wollstonecraft died, and so did the women's right's movement, not to be resurrected for 51 years.

In the United States, seventy-two years after the *Declaration of Independence* was first crafted by Thomas Jefferson, a new document was published called the *Declaration of Sentiments*. It called for woman's suffrage and put new meaning into the *Declaration of Independence*. Its authors argued that since women were equal to men, they should be treated as equals in the law and in political participation.

In Seneca Falls, New York, at the first National Women's Rights Convention in 1848, Elisabeth Cady Stanton launched the National Women's Rights movement in the United States.

Three years later, another equally prominent woman emerged. Susan B. Anthony (1820-1906) was an advocate and activist in the Women's Suffrage Movement. Anthony was the daughter of an affluent cotton manufacturer who was so opposed to slavery, that he limited the purchase of cotton for his mills to plantation owners who did not use slave labor. Anthony inherited his activist characteristics. She taught school until 1846. Her last teaching assignment was as Head of the Female Department of Canajoharie Academy.

By 1851, she became acquainted with Amelia Bloomer, Lucretia Mott, Lucy Stone, and Elisabeth Cady Stanton. The latter became a life long friend and together they would accomplish much for the women's movement.

In 1852, Anthony organized the "Women's State Temperance Society of New York." There was enormous opposition to women participating in public affairs. It was during this period that she was convinced that women would only become effective workers for reform by obtaining equal rights. That same year, she joined Amelia Bloomer and others who dressed in short skirts and Turkish trousers. These garments came to be known as the bloomer costume, and finally simply "Bloomers."

After multiple lecture tours in 1868, Anthony joined with Stanton and Parker Pillsbury in publishing a radical and defiant periodical called *The Revolution*.

The "National Women's Suffrage Association" was organized in 1869 to try to get a sixteenth amendment affixed to the Federal Constitution that would enfranchise women. Mrs. Stanton was the first President. Susan B. Anthony was Chairwoman of the Executive Committee.

That same year, the 14th Amendment to the United States Constitution was passed to protect the civil rights of freedmen. In 1870, the 15th Amendment was ratified. It outlawed disfranchisement "on account of race, color or previous condition of servitude."

In 1872, in Rochester, New York, as a test of the new 14th Amendment, Susan B. Anthony and fifteen other women registered and voted in the November elections. Two weeks later, she was arrested in violation of the election laws. Since her trial was postponed, she defiantly voted again in the March, 1873, city elections.

At her trial (United States vs. Susan B. Anthony), she attempted in vain to have women sit on the jury. Judge Ward Hunt delivered a written opinion, which he had drafted before the trial began. He directed the jury to find the defendant "guilty." The result was that Anthony was fined $100. She told Judge Hunt that she would never pay a dollar of the fine and she kept her word. Instead, she continued to speak out in favor of the improvement of the social, legal and political rights of women. Although progress was slow, events were developing that would make a difference in not only the women's movement, but in the American and international industrial offices of the world.

Time line posters as supplied through the courtesy of the National Women's History Project, 7738 Bell Road, Windsor, California 95492.

Because of the shortage of manpower during the American Civil War, President Abraham Lincoln's Secretary of the Treasury, startled the country by hiring several hundred women to fill vacant Government clerical positions. This was the first time women were hired in large numbers for office clerical positions.

But, even more important was a newly announced invention called the typewriter. Christopher Latham Sholes had invented a machine that would print letters. E. Remington & Sons in Ilion, New York, was to manufacture the machine. Since it had a keyboard, not unlike a piano, it seemed to be custom-made for a woman's hands.

Christopher Latham Sholes' daughter, Lillian, poses for a picture of the Sholes experimental typewriter.

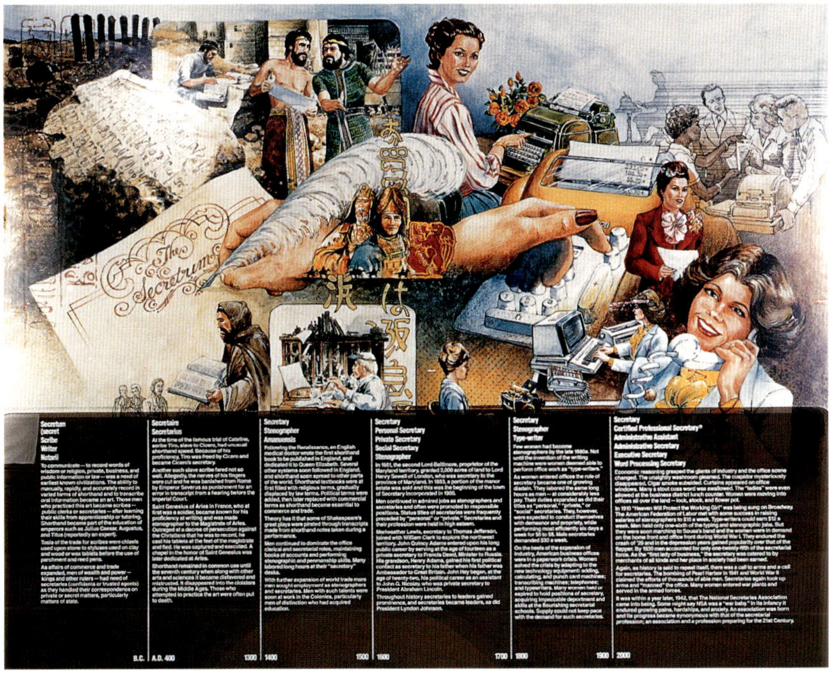

Two charts developed by the Professional Secretaries International organization which were exhibited at the Business Technology Association in Kansas City, Missouri and later acquired by the author. The chart describes the progress of women typists and secretaries in the American offices and how they developed over the years.

Business schools teaching touch typing sprang up all over the country and women excelled. However, the first person to use the touch method of typing was a man, Frank E. McGurrin, in 1878, on the Model 1 Remington. The first business school to begin offering instruction in touch-typing was the Longley Shorthand and Typewriter Institute of Cincinnati, Ohio, which was eventually owned and operated by Mrs. Longley. In 1881, she called her teaching process the "All Finger Method." One year later, Mrs. Longley published her own book called, *Remington Typewriter Lessons*. This was the first printed system for teaching the "All Finger Method." The first person to use the word "touch," in connection with the typing system was Mr. Bates Torrey, of Portland, Maine. According to a Remington article, there was no evidence that the word "touch" as applied to typewriting, was ever used prior to the publication of Mr. Torrey's book, *A Manual of Practical Typewriting*, in 1889.

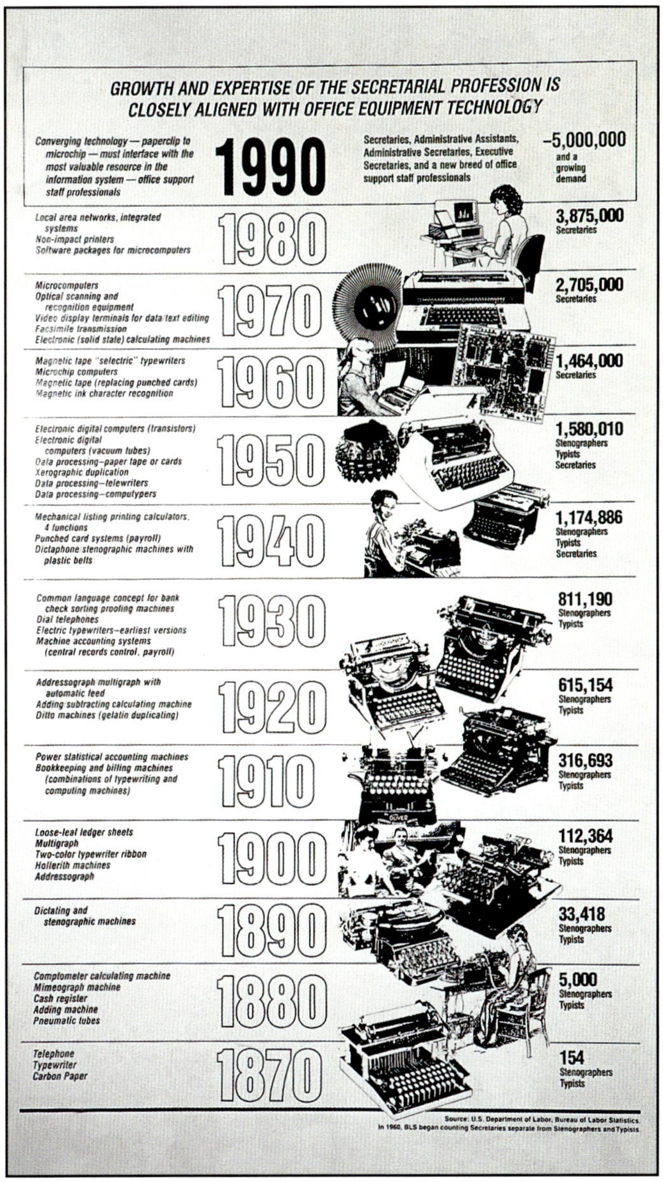

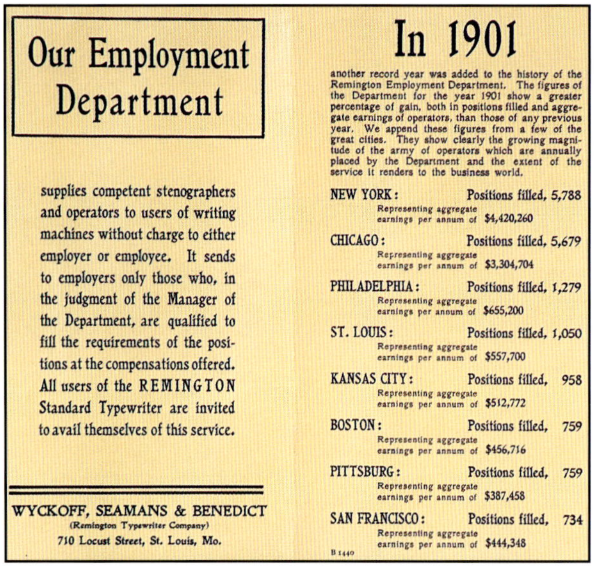

Our Employment Department

supplies competent stenographers and operators to users of writing machines without charge to either employer or employee. It sends to employers only those who, in the judgment of the Manager of the Department, are qualified to fill the requirements of the positions at the compensations offered. All users of the REMINGTON Standard Typewriter are invited to avail themselves of this service.

WYCKOFF, SEAMANS & BENEDICT
(Remington Typewriter Company)
710 Locust Street, St. Louis, Mo.

In 1901

another record year was added to the history of the Remington Employment Department. The figures of the Department for the year 1901 show a greater percentage of gain, both in positions filled and aggregate earnings of operators, than those of any previous year. We append these figures from a few of the great cities. They show clearly the growing magnitude of the army of operators which are annually placed by the Department and the extent of the service it renders to the business world.

City	Positions filled	Representing aggregate earnings per annum of
NEW YORK:	5,788	$4,420,260
CHICAGO:	5,679	$3,304,704
PHILADELPHIA:	1,279	$655,200
ST. LOUIS:	1,050	$557,700
KANSAS CITY:	958	$512,772
BOSTON:	759	$456,716
PITTSBURG:	759	$387,458
SAN FRANCISCO:	734	$444,348

The previously mentioned chart by Professional Secretaries International, tracked women's employment only up to 1890 for a total of 5,000,000 employed. By 1950, that figure escalated to over 22,000,000 women employed.

Clearly, women had arrived in the front office of industrial America and the typewriter had played a major role in that transition.

Picture titled "Comrades" indicating the close relationship between the secretary and her typewriter. Copyright 1903 by the Remington Typewriter Company. Picture enhanced in 2001 by Ms. Beth Trepper, a photographer in Wilmington, Delaware.

Top left, center left & left: Remington organized a "Typist Placement Department" to aid women in finding employment. Also shown: the brochure used and a picture of the waiting room with women waiting for their assignments.

Chapter Four
THE REMINGTON TYPEWRITER COMPANY: WYCOFF, SEAMANS & BENEDICT

After the sale of the Remington Typewriter Division to Wyckoff, Seamans & Benedict, typewriter sales climbed steadily and distribution became global. For several years, the Remington Model 2 was the top performer with accumulated sales approaching 100,000 units.

To meet the demands of the marketplace, three new models were added to the line, Models 3, 4, and 5. According to an 1890 publication, *Manual of the Remington Standard Typewriter* by John Harrison of Pitman & Sons in London, England, and Wyckoff, Seamans & Benedict, the following information describes in detail how they differed from the Remington Model 2. (Since the Models 3 and 5 are of such similar construction, the comparison is made between the Model 2 and 5.)

- rack is reversed so that its teeth point upwards instead of down.
- rack is made to rock instead of the dogs.
- dogs are above the rack and travel along it. This allows the carriage to be pushed from left to right without injuring the rack or dogs.
- carriage is narrower and lighter in proportion.
- key-levers have an upper instead of lower arrangement of springs.
- connecting-wires are straighter, divided and joined by small nuts, thus rendering it easier to shorten or lengthen the wire when occasion demands it.
- four extra keys, writing eight additional characters.
- bell-ringer and the bell are in front, so that they can be seen.
- has a line spacing gauge regulating three widths.
- carriage lever is of a different and simpler construction.

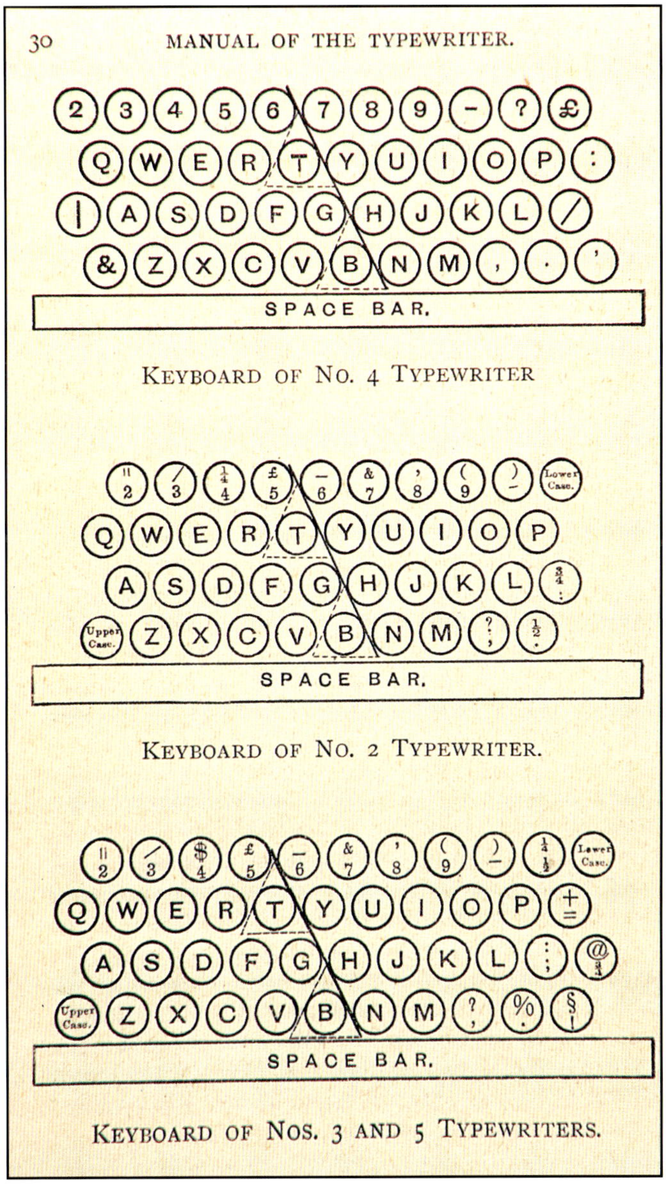

Keyboard arrangements on the Models 2, 3, 4, and 5.

The Remington Model 4 is built upon the style of the Model 2, but has the Model 1 keyboard. It is less expensive than the Model 2, but it has only one set of type, all capitals.

By 1890, all five (5) models of the Remington were being sold with the following recommendations:

- The Model 1 is built upon the original model with modifications. It is cumbersome, and it is not advisable to purchase it except where money is an important item. To use it as means with which to earn a living was useless since it has capitals only. It is, however, a good strong machine and will stand hard work.
- The Model 2 Remington is the popular instrument. It has capitals and small letters, figures, punctuation marks, and the following signs: 1/4, 1/2, 3/4 and £. It will write on paper up to 8" wide and of indefinite length. It is therefore suitable for commercial and general use.
- The Model 3 Remington is adapted for use by lawyers, insurance companies and all needing to write on paper of extra width. It will take paper up to 14" wide. Its keys and characters are similar to the Model 5.

- The Model 4 Remington is built on the style of the Model 2 with a Model 1 keyboard. It is less expensive than the Model 2, but it supports capitals only. It is recommended for those who do their own writing and to whom money is an object.
- The Model 5 is capable of taking a sheet of paper 9-1/2" wide. It has four keys and eight characters in addition to those on the Model 2. This is the latest improvement in typewriting machines and is similar in construction to the Model 3 machine.

During this same year, 1890, sales of the Remington typewriter reached 20,000 machines per annum. In 1892, the ability of the Remington factory to produce 100 machines a day was inadequate to meet the rapidly increasing public demand.

In June 1888, P. H. Hubert, Jr. wrote an article entitled, "The Typewriter: Its Growth and Uses." He reported that "the chief reason for the success and popularity of the typewriter is, of course, the saving of time effected in business offices. This has now become so well known as to scarcely need argument and explains the marvelous increase in the sale of the Remington Typewriters."

Hubert goes on to say that "based on a countless number of tests coupled with the experiences of nine out of ten of our active business houses, the typewriter, as compared to the pen, saves forty minutes an hour." To carry out the calculation that equated to five hours and twenty minutes in a business day.

Think of it, forty minutes an hour. The typewriter was beginning to realize its potential by demonstrating the dramatic savings of time in business offices in the United States or the World. Is it any wonder that Remington's Trademark was changed and was advertised in 84 different languages with the motto, "To Save Time Is To Lengthen Life"?

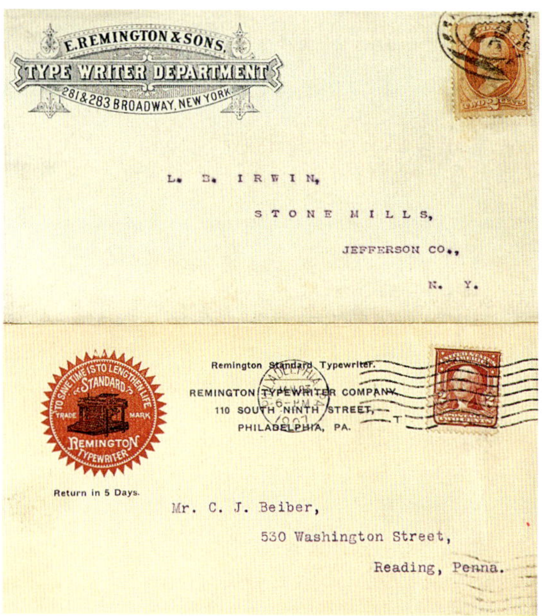

Remington Red Seal Trademark as it appears on an envelop postmarked 1907.

In 1888 Miss Emiline S. Owen, while demonstrating the Remington in England, typed for three minutes at a speed of 99 words per minute. Mr. E.D. Easton, one of the leading legal stenographers of Washington, D.C., said that in one of their famous cases, the Guiteau Case, two operators and machines were used for setting up the copy. Approximately 21,000 folios were transcribed, representing savings over the old method of nearly four cents per folio or $840 in three months.

Miss Mary E. Orr (August 14, 1888)

Miss Mary E. Orr saw an opportunity for women in business with the new typewriter technology and opened a "Copy Office" in New York. She typed and retyped documents to answer the need for file copies and in the process earned as much as $3,000 a year. She had won several typing contests in the past, but never an international tournament until the Convention of the Canadian Shorthand Society held in Toronto, Canada. Miss Orr won the Gold Medal representing the championship of the world. Using a Remington typewriter, she averaged over 100 words per minute.

In June 1889, at a speed exhibition at Mead's Business College, Miss Orr attempted to break the speed record on "new matter," which was 125 words per minute. She proceeded to give what proved to be the best performance of her life by typing 130 words per minute to break the record.

The Remington Typewriter Company hired Miss Orr. Over the years she excelled in her duties and advanced to become a member of the Board of Directors. She was one of the first women to hold such a position in a major American corporation.

In 1973, on the 100th Year Anniversary of the Remington typewriter, young women were still being trained on the typewriter with the original "QWERTY" keyboard and using these skills to compete for positions in industrial offices throughout the nation.

Bronze medal awarded to the author's eldest daughter, Kathleen Russo Ryan in June 1973 when she was honored as the "Business Student of the Year."

The typewriter began to fulfill its original promise to be "Three Times Faster Than The Pen," and then some, if one considers that in 1853, the handwriting speed record was 30 words per minute.

On Thursday, October 4, 1906, the local Ilion, New York, newspaper, *"Ilion Citizen,"* ran a special edition entitled the "Story of an Ideal Plant."

THE RECREATION PARK.

SYSTEM CHART OF THE REMINGTON TYPEWRITER FACTORY.

ILION CITIZEN.

ILION CITIZEN, THURSDAY, OCTOBER 4, 1906.

Special Illustrated Industrial Representation of Ilion.

The Unparalleled Industry of the Gun and Typewriter Town Told in a Graphic Narrative by Picture and Pen--An Interesting Story of an Interesting Town.

A Story Which May Well Hold the Attention of Any Business Seeking Advantageous Location. The Name "Ilion" is a Trade Mark Known in Every Quarter of the Earth as the Synonym of Highest Quality and Absolute Reliability. The Mammoth Factories, Unequalled Merit and World-wide Fame of the Remington Typewriters and the Remington Guns Have Made a Reputation for Ilion Which is All, Without One Dollar's Expense, Placed Behind Any Enterprise Which is so Fortunate as to Make Ilion its Home. This is a Business Asset which can be Secured at First Hand in no Other Town in the Mohawk Valley. Read the Story and Note the Evidence of Prosperity of Ilion, More Widely Known and More Highly Famed than the Ilium of Olden Days.

THE REMINGTON ARMS CO.

THE INDUSTRY THAT MADE ILION FAMOUS.

THE TYPEWRITER WORKS.

ILION'S PREMIER INDUSTRY, THE REMINGTON TYPEWRITER.

Ilion Citizen – October 4, 1906. Two pages devoted to the Remington factory.

The Remington typewriter had been on the market for 33 years and Remington, as well as the city of Ilion, New York was proud of their local factory. A factory that employed over 1,600 local citizens dedicated to the manufacture of the latest technology. This was not an ordinary factory. The Remington system of management as well as manufacturing was an example that other manufacturers could aspire. It was the biggest typewriter plant in the world and was capable of turning out 100,000 typewriters of the current model per year. That was equivalent to over a machine a minute. There were over 500,000 of these current models in circulation and thousands of older models.

In the early days of Wyckoff, Seamans & Benedict, Remington sales hardly exceeded sixty machines a month. In the early 1890s, sales rose to a machine every five minutes. In 1906, the company was booking orders at the previously unheard of rate of sixty machines an hour.

The Remington factory was global in scope. By that, we mean it was shipping machines all over the world from their Ilion, New York plant. They were manufacturing machines in many languages; i.e. Burmese, Armenian, Laos, Arabic, Yiddish, Hebrew, Hindi, Persian, and Japanese in addition to English.

The factory was almost a city within a city. Its capabilities included a brass foundry, an iron foundry, inspection, cleaning and grinding of castings, coal storage, power house, japanning, tin shop, packing box and wood working, rubber work, forging, annealing, and hardening.

Remington management was even more proud of their relationship with their employees, not to mention, with the city of Ilion. This *esprit de corps* did not just happen. Remington management worked hard to develop it. For instance, in the organizational chart of the company was a department called "Industrial Betterment." It included such functions as:

- **Service Badges.** Gold medals awarded for 10, 15, 20, 25, 30, and 35 years of continuous service.
- **Bonus.** $100 gold coin was awarded to the most efficient employees.
- **Employee Suggestions.** A $20 award for a good suggestion.
- **Free Hot and Cold Water Baths.**
- **Seamans Free Library.**
- **A Recreation Park.**
- **A Young Women's Tennis Club.**
- **The Foreman's Cricket Club.**
- **Six Baseball Teams.** Individual departments had their own teams and competed for the coveted annual "Clarence W. Seamans Trophy."
- **A Remington Typewriter Band.**
- **A Typewriter Hose Company.**
- **A First Aid Department.**

Remington Typewriter Company (RTC) Service Medals

Obviously, they offered a square deal to their employees. The employees responded with the respect and loyalty that was the envy of the industry.

The first Remington Portable typewriter made its debut in 1920. It was an overnight success. Remington followed up with a variety of models targeting different applications. The Model 1 Remington Portable had one shift key on the left-hand side of the machine. In order to keep the machine slim for portability, a unique system of folding the type was designed. A lever on the right side of the typewriter moved the type up several degrees into position to type. When the lever was reversed, the type was folded back down into its compact resting position.

In September 1923, the Remington Typewriter Company celebrated the 50th Anniversary of the Typewriter. The scope of the celebration can best be described by the attendance of 6,000 visitors to the city of Ilion, which only had 12,000 inhabitants. The Remington Typewriter Company also invited all of its major competitors to help celebrate its 50th anniversary. A special train was chartered just to bring visitors from New York City to Ilion. As the passengers disembarked the train, they were met and escorted by the thirty-five member Remington Band. Its bandleader was a former member of the famous band of John Phillip Sousa.

Many speeches were made but the highlight of the affair was the forty-five (45) minute address by Henry Harper Benedict, the soul survivor of the original pioneers of the Wyckoff, Seamans & Benedict firm. Recalling the events of the past fifty (50) years was a nostalgic moment for everyone, but the history of the typewriter was only half over.

In January 1924, the Remington Typewriter Company joined forces with the Noiseless Typewriter Company and formed the Remington-Noiseless Typewriter Corporation. They continued to manufacturer noiseless typewriters at the Middletown, Connecticut factory. The first product was the Remington Noiseless Model 5, which entered the market in 1924. It still had a three-bank keyboard with a double shift mechanism.

Remington Noiseless (Model 5) – *Typewriter Topics*, January, 1925 issue.

Remington Noiseless (Model 6) – *Typewriter Topics*, June 1925 Issue.

Remington Electric Typewriter (Model – 1925).

In 1925, Remington introduced its first electric typewriter. It was America's first fully automatic electric typewriter. Utilizing James Smather's electric power mechanism, Remington completely electrified their Model 12. Since it was not readily accepted by the market not many were manufactured.

In 1927, the decision was made not to spend a million dollars to retool the factory for the production of the Model 12. This would prove later to have been a poor decision by Remington Rand. IBM picked up the Smather's patent and became the dominant force behind the electric typewriter.

The second half of the history of the Remington typewriter was launched by James Henry Rand, Sr. (May 29, 1859 – September 15, 1944) an inventor and entrepreneur. When he was 18, Rand, Sr. took a job as a telegraph operator. Shortly after, he became a cashier at a Tonawanda, New York bank. It was here that he realized what a great deal of time was being wasted searching for records. In those days, files did not have indexes or dividers. Rand, Sr. developed what he called a "Visible" index and divider system of filing with colored signal strips and tabs. He formed a firm called the Rand Ledger Company. He invented new systems such as index files for bank deposit tickets and checks, a stop-payment register, sorters, and similar devices. His company developed a nationwide reputation for excellence in the business systems field.

Rand, Sr.'s first son, James Henry Rand Jr., one of five children, entered the family business after graduating Harvard in 1908. He was in charge of his father's company from 1910 to 1914. A disagreement on how to best grow the company caused Rand, Jr. to leave the company. With $10,000 of borrowed money, Rand Jr.

James H. Rand Jr. of Remington Rand

James H. Rand, Jr.

started his own company, the American Kardex Company. Within five years the new company was doing over one million in gross sales per year. At the insistence of his mother, Rand Jr. and Rand Sr. joined forces to create the Rand Kardex Company in 1925. The younger Rand became President and the elder became Chairman. Rand Kardex was then the largest records retrieval company in the United States. Their offerings included over 4,000 products distributed through 4,500 field sales representatives.

Within a year of their reunion, the Rand Kardex Company began a period of mergers and acquisitions that included Library Bureau, which had patented the first vertical filing system in 1882, and the Safe-Cabinet Company, which pioneered fire protection files and safes. But by far, the most significant merger of Rand-Kardex, was with the Remington Typewriter Company. This was announced on February 10, 1927. The new firm was called Remington Rand and had assets of $73,000,000, earning a net profit of $2,800,000 in their first year. During this same period, Baker-Vawter, Powers Tabulating, and the Dalton Adding Machine Company were acquired. The Remington Rand Corporation was then the largest office equipment company in the world.

Remington Rand under the leadership of James H. Rand, Jr., did not enjoy the same good relationship with its factory workers as either the Remington family or Wyckoff, Seamans & Benedict had. On July 31, 1936, violence broke out in Remington Rand's Norwood plant that was just outside of Cincinnati, Ohio. Less than one month later, on August 22, 1936, a riot broke out at the Syracuse, New York plant whose workers were on strike. Several shots were fired. Throughout Remington Rand's history, employee relations were marked with strikes and violence.

After 63 years of continuous typewriter manufacture at the Ilion factory, Remington Rand moved their typewriter production facilities to Elmira, New York. Production was phased in gradually beginning in 1936.

During World War II, most typewriter manufacturers converted production over to military items such as the Norden Bomb Sight. This was also true of Remington Rand who officially discontinued production of typewriters in October 1942. Remington Rand converted most of its production to Hamilton standard propellers, Colt automatic pistols, projectiles, control systems, shell fuses and primers, aircraft rudders, stabilizers, ammunition loading equipment, machine gun parts, torpedo parts, and radio special equipment. Secret electronic weapons were developed in the company laboratories. During this period, typewriters became very scarce. The IBM electric typewriter was about the only machine that was still available. As a result, IBM got a big jump on the market and would later account for approximately 80 percent of it.

Remington Rand reentered the electric typewriter market in July 1948 with the Remington Electric deLuxe. Unfortunately by now, typists were accustomed to the IBM touch and Remington market penetration started out slow and then tapered off. These results proved to be more important than Remington realized at the time. It allowed IBM to draw large profits from their typewriter division to help finance their Data Processing efforts.

Remington Rand advertisement during World War II showing a B-17 bomber with a Remington Model 17 typewriter.

In January 1939, Remington began production of a new standard typewriter design, the Model 17. This model was probably the best typewriter Remington ever built. It had a light, responsive touch, and operator acceptance was excellent. But there was another manual typewriter on the market with a good touch - the Royal typewriter. In addition to a light touch, the Royal had an innovative feature called "Magic Margins." Simply stated the margins were spring driven as opposed to being moved manually.

In June 1946, Remington introduced their Model 17 KMC. The KMC stood for "Keyboard Margin Control." This was a spring driven margin similar to the Royal Typewriter. It was so similar, that Royal filed a patent infringement suit against Remington and won, forcing Remington to take the KMC spring driven margins off their Model 17s.

Remington now found itself faced with fierce competition from IBM in electric typewriters as well as from Royal in manual typewriters.

It was during this same time frame, 1946, two University of Pennsylvania engineers, J. Presper Eckert and John W. Mauchly, developed America's first all-electronic digital computer known as ENIAC (Electronic Numerical Integrator and Calculator). Soon after development of ENIAC, Eckert and Mauchly formed a new company called the Electronic Control Company, later called the Eckert-Mauchly Computer Corporation. Their Univac I was the world's first general-purpose commercial computer and was able to handle a wide variety of applications.

In 1950, Remington Rand acquired the Eckert-Mauchly Computer Corporation. In 1951, Remington Rand sold the first Univac to the United States Census Bureau.

In 1951, General Douglas MacArthur, a well known Five-Star General and hero of World Wars I and II, was relieved of his command during the Korean War by President Harry S. Truman. With his retirement from active duty, MacArthur was immediately asked to run for the President of the United States in the next national election by the Republican Party. He declined to be nominated. Instead, in 1952 accepted a position in the private sector as the Chairman of the Board of Remington Rand.

Also in 1952, Remington Rand acquired the Engineering Research Associates of St. Paul, Minnesota. ERA consisted of a group of mathematicians and engineers who, during the World War II, were active in the development of electronic cryptographic equipment. After the war, ERA developed and delivered the first scientific computers in the United States.

In 1955, Remington Rand, Inc. merged with the Sperry Corporation and formed the Sperry Rand Corporation. Their name was later changed to Sperry Univac.

On August 27, 1974, Sperry Univac merged with another division of Sperry Rand, Remington Office Systems and Machines. By 1975, Sperry Univac enjoyed revenues of $1.3 billion.

While 1973 should have been a time of celebration through advertising because it was the 100th anniversary of the typewriter, Sperry Univac chose to downplay this remarkable period. The Typewriter Division was rumored to be up for sale. It would prove to be a complex sale. The Remington name was part of several different divisions. There was the Office Machines Division, whose products included typewriters, adding machines, calculators, and copiers, as well as consumer products such as low price portable typewriters, adding machines, etc. There was the Remington Office Systems Division, whose products included record retrieval items such as shelf filing, Lektrievers or power files, file folders etc. Finally, there was the Remington Shaver Division who sold shavers, hair dyers, electric knives and other consumer products. They had already sold the Remington Shaver Division and Remington name to Victor Kiam. The Remington Rand name could only be used for consumer products. When the Remington Office Systems Division was sold, it did not retain the Remington name but instead called their company simply Kardex, the name of one of their early visible filing systems. This meant that the Remington Rand Office Machines Division could be sold with the Remington Rand name, which was a big advantage.

In 1978, Bert Williams, a former IBM executive, persuaded a private group of investors to finance the acquisition of the Remington Rand division from Sperry Univac. The price was reported to be between twenty-five and twenty-six million dollars. Williams believed that he could make the Remington name the powerful force that it had once been in the office equipment industry. Over $4 million dollars were spent to retool the factory in order to produce their version of the IBM Selectric typewriter, the SE100. Initially, the new management operated out of Blue Bell, Pennsylvania. Within a short period of time, they moved their operations to 103 College Road in Princeton, New Jersey.

The Remington factory in the Netherlands, about an hour outside of Amsterdam, was given the objective to replicate the IBM Selectric typewriter through reverse engineering.

When the first model, the SE100, was released to the field, their dealer distribution channel enthusiastically received it. But, severe mechanical service problems persisted. The initial introductory momentum was soon lost. The problem was reported to be the improper hardening of metal parts.

By the time the problem was corrected, Remington Rand began to experience serious financial problems. Williams lost the support of his financial backers and was replaced by G. G. Cope.

In 1978, the same year that Remington Rand began to reproduce the IBM Selectric typewriter, Olivetti introduced what was reported to be the world's first commercially successful electronic typewriters, the Models 201 and 221. While Remington Rand was struggling to debug their single element typewriter, the industry was rapidly converting their product line to electronic typewriters with daisy wheel print heads as opposed to the golf ball or single element mechanical machines.

On March 28, 1981, Remington Rand filed for protection and reorganization under Chapter 11 of the Federal Bankruptcy Code. Seven months later, on October 15, 1981, Remington Rand filed a proposed plan of reorganization for the payment of creditors in Federal Court in Trenton, New Jersey.

The Dutch government does not have the same bankruptcy laws as in the United States. As a result, they would not allow the factory in Holland to be shut down and its employees terminated. Instead, the Dutch government allowed the plant, which at that time was owned by Arab investors, to continue manufacturing the single element typewriters under the BSI (Business Systems, Incorporated) logo.

On April 27, 1981, G. G. Cope, Group Vice President of Remington Rand, sent a letter to its distribution channels outlining the reasons for the company's bankruptcy. He cited the following three developments:

- A decline in sales in the European marketplace due to severe recessions, in addition to the impact of the electronic typewriter and the aggressive price reduction by IBM in the international marketplace. It should be noted the sales in the U. S. are 26% ahead of last year.
- The extraordinarily high cost of money, with the prime rate soaring to record levels.
- The impact on cash flow caused by our discontinuance of the copier line and accepting copier returns from our U.S. dealers.

The letter went on to say that "while the company could have absorbed any two of the above circumstances, it was the impact of all three factors happening simultaneously that resulted in the necessity to file for reorganization under Chapter 11."

Cope announced to Remington distribution channels on December 10, 1981, that the company would launch a new line of electronic typewriters. The electronic Models 210 and 220 were announced on December 18, 1981. These electronics were OEM products made in West Germany for Remington Rand.

On September 9, 1981, the following joint statement was released to the field distribution channel:

> "Remington Rand Corporation and Business Systems Incorporated (BSI) have entered into a temporary sales agreement covering the sale of Model "101" typewriters by BSI to Remington Rand.
>
> "This agreement is a temporary arrangement pending the possible confirmation of a plan of Reorganization offered in Remington Rand's pending Chapter 11 case by the Creditors' Committee with the support of BSI"

Remington Rand placed what would prove to be their last national advertising piece in the Wall Street Journal on December 10, 1981, on the electronic typewriter models 210 and 220.

Remington Rand's last national advertisement – *Wall Street Journal* – December 10, 1981. Original advertisement provided by Anthony Casillo of Garden City South, New York.

That same month, on December 28, 1981, Remington Rand sent the following Western Union telegram to their dealers:

A BELATED CHRISTMAS GIFT
WE'VE DONE IT. THANKS TO THE SUPPORT OF OUR DEALERS AND A LOT OF HARD WORK BY THE EMPLOYEES OF REMINGTON RAND, WE ARE OUT OF CHAPTER 11.

THE ORDER APPROVING OUR REORGANIZATION PLAN WAS SIGNED IN FEDERAL COURT IN TRENTON, NEW JERSEY, ON DECEMBER 24, 1981.

WE KNOW THAT THIS WILL HELP US ALL TO MAKE 1982 A HAPPY AND PROSPEROUS NEW YEAR.

ADDITIONAL INFORMATION TO FOLLOW.

Remington Rand Vice President announced, on July 1, 1982, that Remington electronic typewriters had been designated as the official typewriter of the 1982 World's Fair in Knoxville, Tennessee. Remington would be showing three new additions to their electronic typewriter line at the NOMDA convention in Kansas City, Missouri, during the same month.

Just when it appeared that Remington Rand's fortunes began to take a turn for a more positive future, on July 5, 1983, Gerald D. Quill, President of Remington Rand, sent a letter to their dealers. They announced that Remington Rand had executed an agreement giving Wordtronix, Inc. of Minneapolis, Minnesota the world-wide rights to all Remington Rand trademarks. The driving force behind the Wordtronix Company was its President, V. A. Kluesner. He was a former CPT Executive who founded Wordtronix on the concept that there was a need for a simple, inexpensive, user friendly word processing work station using an IBM Selectric typewriter connected to a full page CRT screen using a Z-80 chip. When they showed it at NOMDA, IBM was showing their word processor unit using an Intel 8080 chip that would make the Wordtronix unit obsolete.

In the same month, July 29, 1983, the merger of Wordtronix with Lee Data Corporation was announced. Lee Data manufactured data terminals and Wordtronix manufactured word processing machines. Remington Rand, which for over 100 years manufactured typewriters, provided the name recognition. Wordtronix would be operated as a wholly owned subsidiary. Lee Data provided Wordtronix with a four to five million dollar loan that was not needed, but accepted anyway.

Soon Wordtronix was struggling financially with bank loans that were secured by Lee Data. But it was not serious enough for bankruptcy.

It was at this point that a new player appeared, Kelly Goff from Atlanta. Goff, after reading about the company in the Wall Street Journal, picked up the phone and called the Wordtronix President to ask if he was interested in selling the entire company, which included Lee Data, Wordtronix, and Remington Rand. Goff was in the satellite transmission business. A deal was cut.

Soon after Goff acquired the company, he sold the Remington Rand name to Victor Kiam of the Remington Shaver Company, where it was retired in an effort to avoid future confusion concerning the Remington name. Mr. Kiam subsequently passed away in 2001.

So the Remington Typewriter Company after many successes and adversities finally retired after approximately 114 years. Its legacy is the QWERTY keyboard that is still the input method of choice for modern desktop and portable computers.

Remington Typewriters 1873-1987

Remington Standard (Model 2). Manufactured by E. Remington & Sons of Ilion, New York. At the present time, the Model 2 can still be found at various antique marketplaces and as a result it has a relatively low price. However, it is still very collectible. It was the first machine to introduce a mechanical feature that allowed the operator to use a shift key for upper or lower case, a principle that is still in use today. Two Remington employees patented the original mechanical innovation: Lucian S. Crandall and Byron Brooks. Their idea was to put two characters on each type lug and use a shift key to move the platen in position so that an upper or lower case character would strike the platen. (c.1883) Serial number: 83,035. Availability: Scarce. Value: $500.

Remington Standard (Model 2). This is another Model 2, but has the Jayne palm shift patent applied for. It used a palm shift on both sides of the keyboard. Royal adopted a similar feature with their HH models in the 1950s. The machine pictured here has a broken frame. Serial number: 53,144. Availability: Only machine known to exist. Value: $ 500.

Remington Standard (Model 3).

Remington Standard (Model 4). Manufactured by E. Remington & Sons in Ilion, New York. Although no factory serial numbers are available for the Model 4, an Ilion factory letter dated January 24, 1930, stated that it was introduced in January 1883. The Model 4 has a larger diameter platen (2.50") similar to the Remington Model 1. Platen length: 8.25" with 44 keys, capital letters only. Serial number: 5,118. Availability: Rare. Value: $1,000.

Remington Standard (Model 5). Manufactured by Wycoff, Seamans & Benedict at Ilion, New York. The January 24, 1930 Remington factory letter lists the Model 5 as first appearing in May 1887 and becoming obsolete in February 1893. Platen length: 9.25" with 42 keys and a platen diameter of 2". It had a shift for upper and lower case. Serial number: 19,154. Availability: Rare. Value: $800.

Remington Standard (Model 6). Manufactured by Wycoff, Seamans, & Benedict. Platen length: 9.00" with 38 keys. Serial number: 139,977. (c.1894) Availability: Common. Value: $75.

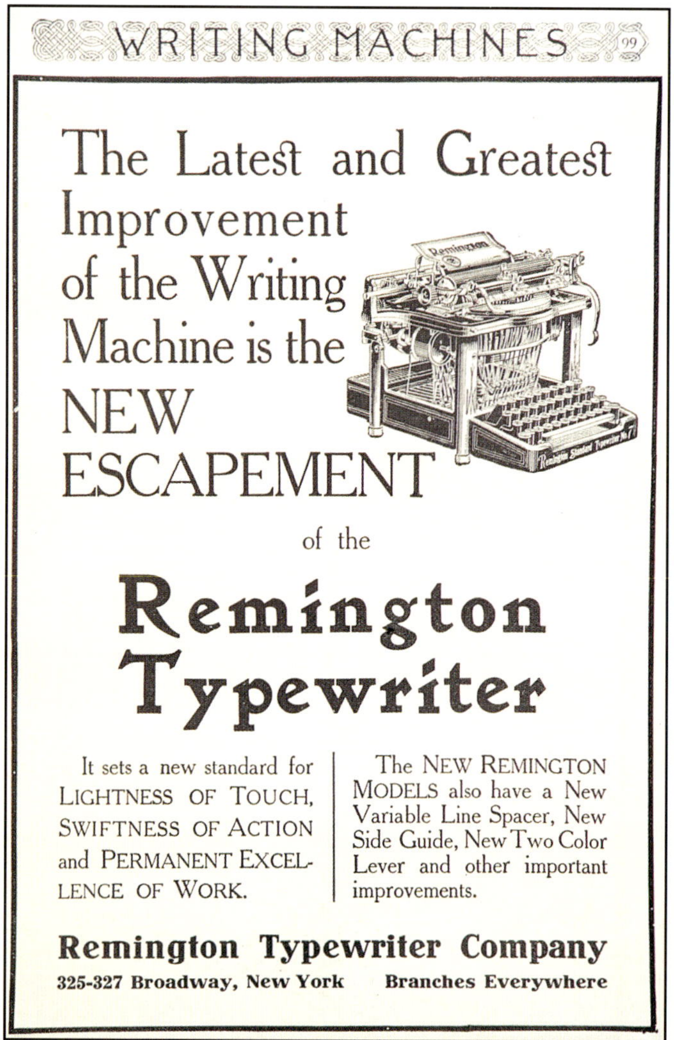

Bottom left, top right and above: **Remington Standard (Model 7).** Manufactured by Wycoff, Seamans, & Benedict. Platen length: 9.50" with 42 keys. Serial number: 204,154. (c.1898) Availability: Common. Value: $75.

Remington Standard (Model 7 – TKT). Manufactured by Wycoff, Seamans & Benedict. This was their first ten-key decimal tabulator machine. It has a plate in the back of the machine mounted on the rear TKT (ten-key tabulator) mechanism that says number 10, Gorin Tabulator, patented and manufactured in Ilion. Platen length: 9.00" with 42 keys. Serial number: 167,775. (c.1905) Availability: Scarce. Value: $200.

Remington Standard (Model 7 – 7KT). Manufactured by Wycoff, Seamans & Benedict. This Remington was historically significant in that it was the actual machine used as a court exhibit during the legal proceeding over patent infringement rights between Underwood and Remington. It has the original tag on it that says, "Defendant's machine in Equity," "Schisinger 7-key Tabulator." It was originally housed in the Remington Museum. Platen length: 9.00" with 42 keys. Serial number: 25,807 (c.1911) Availability: One of a kind. Value: $2,000.

Remington Standard (Model 8). Manufactured by Wycoff, Seamans & Benedict. First introduced in October 1897. Platen length: 9.25" with 42 keys. Carriage: Size A. Serial number: 25,335 (c.1904) Availability: Scarce. Value: $350.

Remington Standard (Model 9 - German). Manufactured by Wycoff, Seamans & Benedict. First introduced in November 1902. From the Remington Museum. German keyboard. Platen length: 9.3125" with 46 keys. Carriage: Size A or 16." Serial number: 15,931. (c.1902) Availability: Scarce. Value: $500.

Remington Standard (Model 9A - Devanagari). Manufactured by Wycoff, Seamans & Benedict. This foreign language machine also came from the Remington museum. Carriage size: A. Platen length: 9.25" with 46 keys. Serial number: 17,191 (c.1907). Availability: Rare. Value: $400.

Remington Standard (Model 9A - Gujarati). Manufactured by Wycoff, Seamans & Benedict. From the Remington Museum with original tags. Platen length: 9.25" with 46 keys. Serial number: 17,146. Availability: Rare. Value: $500.

Remington Standard (Model 9 – German - TKT). Manufactured by Wycoff, Seamans & Benedict. A historically significant machine, it came from Remington Museum and probably was the first production model with no serial number. Platen length: 13.50"and with 45 keys. German Keyboard. Goren tabulator number 10. Availability: Rare. Value: $500.

Remington Standard (Model 92B – Hebrew). Manufactured by Wycoff, Seamans & Benedict. From the Remington Museum with original tags. Platen length: 11.50"with 42 keys. Serial number: 62,921. (c.1931) Availability: Rare. Value: $500.

Remington Standard (Model 9 - Japanese). Manufactured by Wycoff, Seaman & Benedict. A historically significant machine since it came from the Remington Museum. Japanese keyboard. Serial number: 17,197. (c.1904) Availability: Scarce. Value: $400.

Remington Standard (Model 10). Manufactured by Wycoff, Seaman & Benedict. Platen length: 9.375" with 42 keys and a 5-key tabulator. Serial number 429,719 (c.1914) Availability: Common. Value: $50.

Remington Standard (Model 10). This machine was designed for the Postal Telegraph Company. It must have been a fairly large order. The Remington name was cast into the metal on the right frame pillar, "Postal Telegraph." Platen length: 9.875" with 42 keys. Serial number: LD50,663 (c.1925) Availability: Rare. Value: $400.

Remington Standard (Model 11 - TKT). Manufactured by Wycoff, Seamans & Benedict. Remington uses the Wahl Register attachment in conjunction with its Ten-Key Decimal Tabulator in the lower position to form what they now call their Accounting Machine. Platen length: 11.375" with 42 keys. Serial number: 42,414 and RX30,179 respectively. (c.1923) Availability: Common. Value: $125.

Remington Standard (Model 11). Ten-Key Tabulator keys in the top row position. Platen length: 13.75" with 42 keys and 5 starter keys. Serial number: 321,525 (c.1913) Availability: Common. Value: $50.

50th Anniversary *of the* Typewriter

1873

Model 1 Remington. The first practical typewriter and the ancestor of all the writing machines in the world today

Cut from the first typewriter catalogue (1874). This picture of the girl at the typewriter was prophetic of the millions of women who have since earned their living through the writing machine

Christopher Latham Sholes, inventor of the typewriter, at work on one of his experimental machines (1872)

FIFTY YEARS AGO, in September, 1873, at the Remington Works, Ilion, N. Y., began the manufacture of a new machine which was destined to revolutionize business, free the world from pen slavery, and complete the economic emancipation of womankind. No other machine, no other invention, no other article of commerce of any kind has ever played a more commanding rôle in the shaping of business and social destiny.

Christopher Latham Sholes was the inventor of this machine, Ilion, N. Y., was its birthplace, and Philo Remington was its first manufacturer. Originally known simply as "*The Type-writer*," it is now known to fame and history as the Remington Model 1.

We are proud of our record as the founders of this great industry. We are equally proud of our many contributions to typewriter development, which have been continuous throughout the entire fifty years of typewriter history.

The first shift-key typewriter, writing both capitals and small letters (1878), was a Remington.

The first typewriter equipped with an automatic ribbon reverse (1896) was a Remington.

The first typewriter equipped with a decimal tabulator (1898) was a Remington.

The first adding and subtracting typewriter appeared in 1907. It was a Remington.

The Remington brought out the first automatic line indenting mechanism (1908), the first key-set decimal tabulator (1911), the complete accounting machine (1914), and the first portable typewriter with complete, four-row standard keyboard (1920).

Here is a record of progressive pioneering unparalleled in the typewriter industry—but even this does not complete the Remington story.

The latest and greatest Remington contribution to typewriter progress is the new Quiet 12. This new Remington is quiet in its operation. The "natural touch" makes its use a pleasure to the operator. The completely enclosed frame is a protection against dust and dirt, and helps to keep the machine in good working condition. And refinements in the printing mechanism insure good work—always.

This latest Remington is a fitting product of the comprehensive and unrivaled experience of its builders. It has the sum of every quality any user has ever wanted in a writing machine.

1923

Model 12 Remington The culmination of 50 years of typewriter progress

The Quiet 12

Remington Typewriter Company

374 Broadway, New York ∾ *Branches Everywhere*

Remington Standard (Model 12). Platen length: 9.875" with 42 keys and a single red key tabulator. Serial number: LM70,572 (c.1927) Availability: Common. Value: $50.

Top left & above: **Remington Standard (Model 12).** Platen length: 10.00" with 42 keys and 5 starter keys. Serial number: 303,541 (c.1923) Availability: Common. Value: $50.

Remington Standard (Model 16). Platen length: 10.625" with 42 keys and a 5-key tabulator. Serial number: 2,448,775 (c.1934) Availability: Common. Value: $50.

Remington Standard (Model 20 TKT). Ten-Key Decimal Tabulator with a continuous paper roll attachment. A very unusual specimen. Remington put the numerals in the lower row of keys just below the space bar. Their logic was sound because when doing a lot of statistical typing, the numbers should be close to the decimal tabulator. Serial number: LM74,589 (c.1927) Availability: Scarce. Value: $75.

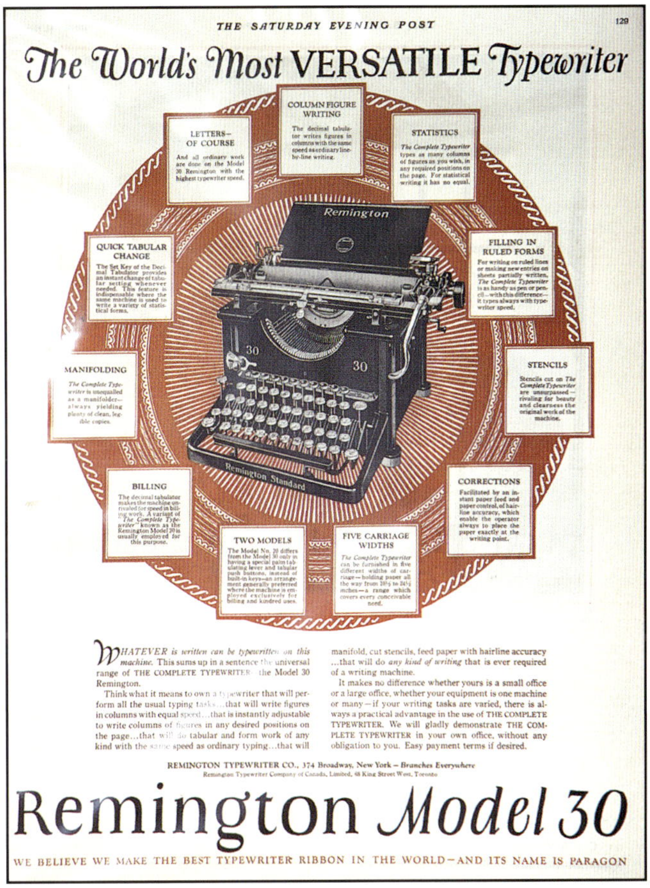

Remington Standard (Model 21 Vertical Adder). Includes ten-key tabulator, palm tabulator, and Wahl adding and subtracting attachment. Platen length: 11.50" with 42 keys. Serial number: 83,019. Availability: Scarce. Value: $250.

Remington Standard (Model 30). With ten-key decimal tabulator on top row of keys. Platen length: 14.00" with 42 keys. Serial number: Z183,004. (c.1929). Availability: Common. Value: $75.

Remington Standard (Model 50). This was an exceptionally long carriage in which the platen measures 22.50" with 42 keys. Because of the size of the carriage, the draw band rewind mechanism was located on the outside of the machine. I was told that this machine came from the garage of the factory superintendent. It was probably an experimental model as I have only seen 20" carriage machines in the past. Serial number: WT148,094. (c.1940) Availability: Rare. Value: $100.

Remington Standard (Model - Monarch). Platen length: 10.25" with 42 keys. Serial number: MC60,216. (c.1916) Availability: Common. Value: $50.

Remington Standard (Model 60). Platen length: 9.375" and double keyboard. Serial number: 132,677. Availability: Common. Value: $125.

100 THE SATURDAY EVENING POST June 24, 1944

B-17
Flying Fortress

MODEL 17
Remington Rand Typewriter

Two famous 17s...both rugged, both leaders

YES, the famous Flying Fortress (B-17) and America's most popular typewriter, the famous Remington Rand (Model 17) are in there pitching in combat zones around the globe. Both have proved that they can take it, and both have seen plenty of action in this war.

For the B-17 we are producing Hamilton Standard Propellers by the thousands, and our deliveries to the Army Air Forces have for many months met all the rigid requirements of quality and quantity. Indeed, on every one of our dozens of war production contracts, deliveries are up to and beyond schedule, and costs to the armed services are uniformly low.

With our war front production thus rolling along full blast, what more natural than that the War Production Board should ask us to turn our attention to a critical need on the Business front?

That's why Remington Rand (Model 17) Typewriters are today being produced in limited quantities to meet only the acute requirements of war and the most essential civilian needs. But the supply is extremely limited—far short of the demand...

...so if you are the lucky owner or operator of a Remington Rand Typewriter, give it the care this indispensable office "machine-tool" really deserves.

Remington Rand

Hamilton Standard Propellers ★ Adding, Calculating and Bookkeeping Machines ★ Colt Automatic Pistols ★ Punched-Card Accounting and Tabulating Machines ★ Projectiles ★ Kardex Visible Control Systems ★ Shell Fuzes and Primers ★ Filing Cabinets, Methods and Supplies ★ Aircraft Rudders and Stabilizers ★ Safe-Cabinets and Safe-Files ★ Ammunition Loading ★ Loose Leaf Systems ★ Machine Gun Parts ★ Photo-Copy and Micro-Film Equipment ★ Torpedo Parts ★ Library Furniture ★ Standard and Noiseless Typewriters ★ Electric Shavers ★ Radio Special Equipment ★ Business Machine Supplies and Nation-wide Maintenance Service ★ Secret Electronic Weapons Developed in the Company's Laboratories.

Remington Standard (Model 17). Removable top-cover. Platen length: 12.75" with 42 keys. Serial number: JC464,265. (c.1942) Availability: Common. Value: $50.

Remington Standard (Model 17 DeLuxe). This was an early Model 17 in which there was a ribbon cover for each ribbon that could be individually lifted, as opposed to a one-piece removable top cover. Platen length: 10.75" with 42 keys. Margins are set manually behind the paper table. Serial number: 5,159. (c.1939) Availability: Scarce. Value: $100.

Remington Standard (Model 17KMC). Remington added an automatic keyboard margin control to this model. It was a spring-loaded margin that was key set, very similar to the Royal Magic Margin set. It proved to be too similar. Royal sued and Remington removed this feature on a later model. Platen length: 10.75" with 42 keys. Serial number: J1,211,972. Availability: Common. Value: $50.

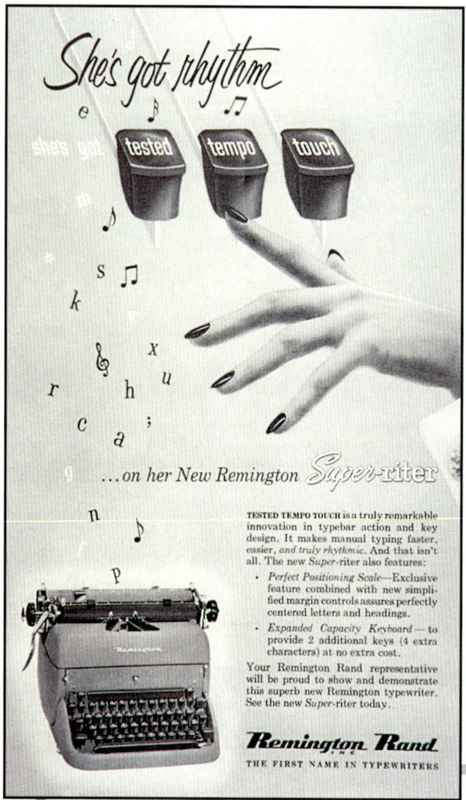

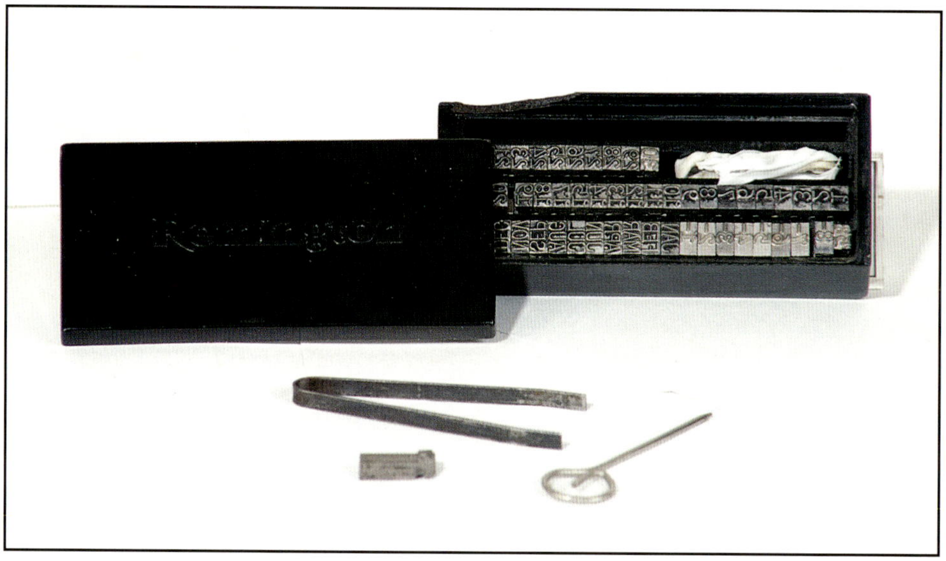

Remington Inter-changeable Type Set. An option Remington introduced during this period that allowed an operator to remove one special type lug from the type basket. Using the tweezers provided, the operator could then remove the typeface and replace it with a special character and/or symbol. Then the process was reversed after the character/symbol was struck. It was also available on the Remington Electric Typewriter. Availability: Scarce. Value complete set: $50.

Remington Standard (Model Super-riter). A total new design change. Platen length: 10.75" with 42 keys. (c.1950). Availability: Common. Value: $25.

Remington Standard (Model 18). Part of a large order for Western Union. An improved model with manual margins. Utilized what was called a "perfect positioning scale" for margins. Instead of starting at zero on the left side of the scale and working to the right, which required margin mathematics, Remington put zero in the center of the scale and worked out both sides. To get a perfectly balanced letter, at whatever position you set the left margin, you would also set the right margin. Platen length: 11.00" with 42 keys. Serial number: J2,671,862. Identified by Western Union as #3513-56, which meant that it was the 56th machine on a multiple order for 3,513 units. Availability: Scarce. Value: $200.

Remington Standard (Model 19). Believed to be the first appearance of a typewriter with a plastic-like casing. The material was called Cycolac, a Borg-Warner product composed of an ABS compound (Acrillian, Budadyene and Styrene). Crash helmets, wedges for splitting logs, and other high impact items were made of Cycolac. It was actually stronger than the traditional die cast aluminum frames. Another added feature is that the color was impregnated into the material eliminating the need to repaint. Carriage length: 13.00" with 44 keys. Serial number: J2,110,296. Availability: Common. Value: $50.

Remington Standard (Model 24). Redesigned case with 13.00" platen and 44 keys. Serial number: M264,228. Availability: Common. Value: $50.

Remington Standard (Model 150).
Platen length: 13.00" with 46 keys. Serial number: S333,539. Availability: Common. Value: $75.

Remington Noiseless Standard (Model 5).
Remington Three-Bank Noiseless. Platen length: 10.25" with 30 keys and double shift. Serial number: 79,033. (c.1924) Availability: Rare. Value: $350.

Remington Noiseless Standard (Model 6).
Remington Rand Four-Bank Noiseless. Platen length: 10.50" with 42 keys. Serial number: X156,571. (c.1929) Availability: Common. Value: $75.

Remington Noiseless Standard (Model 10). Platen length: 10.875" with 42 keys and a tab set and clear key. Serial number: X570,802. (c.1941) Availability: Common. Value: $50.

Remington Noiseless Standard (Model 11). The Remington name is embossed on the paper table. Platen length: 12.25" with 42 keys. Serial number: T25,759. (c.1936) Availability: Scarce. Value: $75.

Remington Standard Noiseless. New redesigned case in a color called Desert Sage. Platen length: 10.625" with 42 keys. Serial number: X818,610. Availability: Common. Value: $50.

Remington Portables

Remington Portable (Model - Junior). This was a three-bank keyboard that appeared in April 1914. Remington discontinued it in 1921. After Remington gave it up, the American Writing Machine Company sold it as the Century. Ribbons are located in the rear of the machine. Platen length: 9.75" with 28 keys and a double shift. Serial number: TP50,338. (c.1914) Availability: Scarce. Value: $200.

Remington Portable (Model 1). The Remington Portable Model 1 was first released in 1920. It had only one shift key, a backspace and pop-up type bars. Platen length: 8.25" with 42 keys. This particular platen has a label holding device. When a small button on the right side of the platen is depressed, two wire hangers are activated for labels. It was a very popular machine for pharmacies because of the size of the machine and the label attachment for drug bottles, etc. Serial number: NA00,261. (c1920). Very low serial number. Availability: Scarce. Value: $200.

Black is Out, Color is In

Remington Portable (Model - Green). First rigid type basket. Serial number: V331,269. (c.1928) Availability: Common. Value: $75.

Remington Portable (Model - Red). Compact model with folding type bars. Platen length: 9.25" with 42 keys. Serial number: NL87,298. (c.1928) Availability: Common. Value: $75.

Remington Portable (Model - Purple). Compact model with folding type bars. Platen length: 9.375 with 43 keys. Tab key, backspace and shift on both sides. Serial number: V249,148. (c.1928) Availability: Common. Value: $75.

Remington Portable (Model – Two-tone Grey and Beige). Platen length: 9.375" with 42 keys and a tab key. Serial number: V348,354. (c.1930) Availability: Common. Value: $75.

Remie Scout

The Remington Remie Scout was a stripped down version of the 1920 compact folding type bars with very few features. It was developed for the educational market and economy was the primary objective. There were three basic models of the Remie Scout, plus a model that was made for Macy's Department Store chain. Remington Rand manufactured approximately 52,800 of these models in the three years between 1932 and 1935. These figures do not include the Monarch Pioneer.

Remie Scout (Model - Single Case). Platen length: 9.375" with 42 keys, pop up type bars and no shift or tab keys. Serial number: S14,718. (c.1932) Availability: Common. Value: $100.

Remie Scout (Model - R. H. Macy Department Store Chain). Single case, no shift or tab key. Serial number: 32,771. (c.1934) Star on left side of top cover and stenciled on rear back plate is "Made for R. H. Macy Co. by Remington Rand." Availability: Scarce. Value: $100.

Remie Scout (Model - Monarch Pioneer). Red and blue, single case, no tab key. Platen length: 9.375" with 42 keys. Serial number: 14,718. (c.1932) Availability: Scarce. Value: $150.

Remie Scout (Model - Streamlined Frame). This model had both upper and lower case. Platen length: 9.375" with 42 keys and no tab. Serial number: S74,611. (c.1934) Availability: Scarce. Value: $75.

Remington Noiseless Portable (Model 7). Pressure printing portable that first appeared on the market in 1931. This was presented as a lightweight desk model. It had many of the features of a standard desk model. Platen length: 9.125" with 42 keys. Serial number: H29,012. Availability: Common. Value: $50.

Remington Noiseless Portable (Model 9). Pressure printing portable that first appeared on the market in 1933. Both the Noiseless Portable Model 8 and the Model 9 were developed to bring the advantage of speed to the small office. Usually these small offices did not warrant the purchase the larger standard desk model. They were exactly alike except that the Model 8 is noiseless and the Model 9, also a noiseless, was referred to by Remington as the "noisey noiseless." Platen length: 10.75" with 42 keys. Serial number: G12,762. (c.1935) Availability: Scarce. Value: $200.

Remington Noiseless Portable (Model 8). Pressure printing portable that first appeared on the market in 1932. Platen length: 10.6875" with 42 keys. Serial number: E15,515. (c.1935) Availability: Common. Value: $50.

Remington Quiet Portable (Model 1). First appeared on the market in 1940. Serial number: P1,167,126. Availability: Common. Value: $75.

Porto-Rite Portable (Model – Two-tone Green). Made for the Sears Roebuck Department Store from 1932 to 1934. Platen length 9.3125" with 42 keys plus a backspace and tab keys. Serial number: ST6,001. (c.1932) Availability: Common. Value: $50.

Remington Portable (Model 5). This portable had a red paragraph key on the right side of the machine. On a later model called the "5T," the paragraph key was changed to a tab key. Platen length: 9.0625" with 43 keys. Serial number: V631,857. (c.1932) Availability: Common. Value: $50.

Remington Portable (Model 5T). Platen 9.00" with 42 keys and tab key. Serial number: V525,117. (c.1932) Availability: Common. Value: $50.

Remington Portable (Model - Home 3B). Factory production started in July 1935 and the product was announced on August 23. It was discontinued in February 1936. Only 5,065 machines were made. This will be a very collectible typewriter in the future, not only because of the small number of machines manufactured, but also because of its unorthodox keyboard. Designated as the #3268 keyboard, Remington eliminated several keys by making the operator shift for several numeral keys. With only 36 keys, the Home 3B had 6 less keys than the average 42-key keyboard. Platen length: 9.3125". Serial number: C103,995. (c.1935) Availability: Scarce. Value: $300.

Remington Portable (Model - Monarch Pioneer). Introduced in October 1937 with the #3268 keyboard. Manufactured for less than a year before discontinuation. Only 17,198 machines were produced. Platen length: 9.25" with 37 keys. Serial number: C117,141. (c.1937) Availability: Scarce. Value: $300.

Remington Portable (Model - Cadet 4B). It did not have margin stops, backspace, tab, or shift lock keys. Platen length: 9.3125" with upper and lower case. Serial number: CB176,949. (c.1938) Availability: Common. Value: $75.

Remington Portable (Model 5 - Streamliner). Announced in November 1935. Platen length: 9.00" with 42 keys and self-starter key. Serial number: V967,550. (c.1935) Availability: Common. Value: $50.

Remington Portable (Model - Envoy). The Envoy has no tab key, but it does have a backspace. Platen length: 9.125" with 42 keys. Serial number: S1,193,163. (c.1938) Availability: Common. Value: $50.

Remington Portable (Model - Remette). Platen length: 9.3125" with 42 keys and no tab key. Serial number: CR296,846. (c.1938) Availability: Common. Value: $50.

Remington Portable (Model - Remette DeLuxe). Backspace key, but no tab key. Platen length: 9.3125" with 42 keys. Serial number: C1,136,340. (c.1940) Availability: Common. Value: $50.

Remington Portable (Model - 1939 World's Fair). Serial number: CR203,667. Availability: Scarce. Value: $600.

Remington Portable (Model - Junior DeLuxe). Platen length: 9.0625" with 42 keys. Serial number: SD277,703. (c.1939) Availability: Common. Value: $50.

Remington Portable (Model - Bantam). Upper case only. Platen length: 9.3125" with 29 keys. Serial number: C130,995. (c.1938) Availability: Common. Value: $50.

Remington Portable (Model 5 - DeLuxe). Has self-starter key. Platen length: 9.125" with 42 keys. Serial number: B1,266,303. (c.1940) Availability: Common. Value: $50.

Remington Portable (Model - Streamliner). Platen length: 9.0625" with 42 keys with a self-starter key. Serial number: B112,196. (c.1941) Availability: Common. Value: $50.

Remington Portable (DeLuxe). Platen length: 9.0125" with 42 keys and a tab key. Serial number: BT1,659,334. (c.1945) Availability: Common. Value: $50.

Remington Portable (Model - All New Remington Portable Typewriter). Manufacture started in June 1949. Platen length: 9.25" with 42 keys. Serial number: AT2,076,819. (c.1949) The model designated with AT indicated that the machine had miracle tabulation, a lever-like device allowing an operator to set or clear a tab. Availability: Common. Value: $50.

Remington Portable (Model - All New Remington Portable Typewriter). This model did not have tabulation. Serial number: A1,991,362. (c.1949) Availability: Common. Value: $50.

Remington Portable (Model - 1949). First announced December 1, 1948. Platen length: 9.125" with 42 keys. Equipped with a self-starter key, but no tab key. Serial number: B1,711,161. (c.1949) Availability: Scarce. Value: $150.

Remington Portable (Model - Office-Riter). The Office-Riter was a portable that had a platen of 10.75" and therefore considered adaptable to the office. It had 44 keys and miracle tabulation. Serial number: ER3,437,643. (c.1953) Availability: Common. Value: $50.

Remington Portable (Model - Quiet-Riter). Equipped with miracle tab; margin release, and touch control. Platen length: 9.4375" with 42 keys. Serial number: QR2,844,296. (c.1953) Availability: Common. Value: $50.

Remington Portable(Model - Quiet-Riter Eleven). Equipped with miracle tab, touch control and margin release key. Platen length: 10.75" with 44 keys. Case is green die cast aluminum. Serial number: QR3,722,273. (c.1956) Availability: Common. Value: $50.

Remington Portable (Model - Quiet-Riter Eleven). With newly designed blue Cycolac case. Platen length: 10.75" with 44 keys and miracle tab and touch control. Serial number: UO4,006,326. (c.1957) Availability: Common. Value: $50.

Remington Portable (Model - Quiet-Riter). Salmon color in Cycolac case. Platen length: 10.4375" with 44 keys. Serial number: UO4,025,555. (c.1959) Availability: Scarce. Value: $125.

Remington Portable (Model - Quiet-Riter). Yellow Cycolac case. Platen length: 9.125" with 42 keys and tab set and clear on wheel. Made in Holland by Sperry Rand. Serial number: LY596,413. (c. 1960) Availability: Common. Value: $50.

Remington Portable (Model - Letter-Riter). Platen length: 9.50" with die cast aluminum frame. No tab key, margin release or touch control. Serial number: LR2,811,320. (c.1955) Availability: Common. Value: $50.

Remington Portable (Model - Letter-Riter DeLuxe). Platen length: 9.50" with 42 keys. Serial number: LT3,623,425. (c.1959) Availability: Scarce. Value: $100.

Remington Portable (Model - Letter-Riter DeLuxe with Cycolac case). Platen length: 9.4375" with 42 keys. Availability: Common. Value: $50.

Remington Portable (Model - Travel-Riter with Blue Cycolac case). Platen length: 9.4375" with 42 keys. No tab key, no margin release, no touch control. Serial number: 701,993D. (c.1959) Availability: Common. Value: $50.

Remington Portable (Model - Fleetwing). Made in Germany. Platen length: 10.50" with 44 keys. Equipped with tab key, margin release, touch control, card holding plastic shields and page end indicator. Serial number: ERF100,749. Availability: Scarce. Value: $75.

Remington Portable (Model - Ten Forty) Made in Holland. Turquoise Cycolac case. Platen length: 9.50" with 43 keys. Serial number: FY576,412. (c.1959) Availability: Common. Value: $25.

Remington Portable (Model - Mark II). Made in Germany. Platen length: 9.375" with 44 keys. Serial number: 832,893. (c.1958) Availability: Scarce. Value: $75.

Remington Portable (Model - Personal-Riter). Blue Cycolac machine made in Holland. Platen length: 9.375" with 43 keys. Serial number: HY664,299. (c.1970) Availability: Common. Value: $25.

Remington Portable (Model - Holiday). Made in Holland. Platen length: 9.50" with 42 keys. Serial number: M367,144. Availability: Common. Value: $25.

Remington Portable (Model - Performer). Made in Japan by Brother. First repeat space bar on a manual typewriter. Platen length: 9.375" with 44 keys. Serial number: J1,689,512. (c.1970) Availability: Scarce. Value: $50.

Remington Portable (Model - Rem-Riter). Platen length: 9.4375" with 42 keys. Serial number: RR144,425. Availability: Scarce. Value: $50.

Remington Portable (Model - Monarch II). Made in Holland. Black and ivory Cycolac case. Platen length: 9.8125" with 42 keys. Serial number: 1X313,559. (c.1970) Availability: Common. Value: $25.

Remington Rand Portable (Model 1). This was the first new portable made after James Rand purchased Remington. Platen length: 9.125" with 43 keys. Serial number: P65,254. (c.1933) Availability: Common. Value: $75.

Remington Portable (Model - Quiet Model 1). Platen length: 9.0625" H with 43 keys. Serial number: P1,115,216. (c.1930) Availability: Scarce. Value: $50.

Remington Portable (Model - Travel-Riter). Made in Holland. Platen length: 9.4375" with 42 keys. Serial number: TR118,672. Availability: Scarce. Value: $50.

Remington Portable (Model - Streamliner Blue Cycolac). Platen length: 9.125" with 42 keys. Serial number: CX264,748. (c. 1960) Availability: Common. Value: $25.

Remington Portable (Model 333). Made in Japan by Brother for Sperry Rand. No tab key. Platen length: 9.375" with 44 keys. Serial number: 7,810,220. (c.1960) Availability: Common. Value: $50.

Remington Portable (Model 666). Made in Japan by Brother for Sperry Rand. Tab set and clear key. Platen length: 9.375" with 44 keys. Serial number: F8,993,521. (c.1960) Availability: Common. Value: $50.

Remington Portable (Model 1 - Premier). Platen length: 9.25" with 42 keys. Serial number: P156,047. (c.1933) Availability: Scarce. Value: $75.

Remington Portable (Model - Noiseless). Above are examples of both the black and two-tone green models. Serial numbers: N38,567 and N67,932, respectively. (c.1935) Availability: Common. Value Each: $50.

Remington Portable (Rem-Blick). After the Blickensderfer patents expired, Remington produced the Rem-Blick. Platen length: 9.75" with 28 Keys and a double shift. Serial number: K080,583. (c.1928) Availability: Scarce. Value: $300.

Remington Portable Electric (Model 611). Manufactured by Brother in Japan. The carriage return was manual, but the keys were electric with an automatic repeat space bar. Platen length: 10.75" with 44 keys. Serial number: BO482,797. (c.1971) Availability: Common. Value: $125.

Remington Electric (Model - 1925). This was Remington's first electric typewriter and was produced under license agreement. However, Remington decided not to renew the agreement and discontinued manufacture after making only a few. IBM picked up the rights and manufactured the machine very successfully for many years. Platen length: 11.5" with 42 keys. Serial number: X3,753. (1925) Availability: Scarce. Value: $1,000.

Remington Portable Electric (Model 612). Manufactured by Brother in Japan. This model had an electric carriage return as well as electric keys. Platen length: 9.75" with 44 keys. Serial number: D1,297,783. (c.1971). Availability: Scarce. Value: $125.

Remington Electric (Model - Electro-conomy). After World War II, Remington developed their own electric typewriter. Platen length: 11.75" with 42 keys. Serial number: E2,187,449. (c.1953) Availability: Scarce. Value: $250.

Remington Electric (Model 250). This model had both fabric and carbon paper ribbons. Platen length: 11.75" with 42 keys. Serial number: E2,187,449. (c.1956) Availability: Common. Value: $125.

Remington Electric (Model 25). Released during the 1960s. It was called the Model 25 because Remington boasted that it had 25 exclusive features. Platen length: 11.75" with 44 keys. Serial number: E542,888. Availability: Common. Value: $125.

Remington Electric (Model 300). Picture taken from a brochure. (c.1961)

Remington Electric (Model 26). Picture taken from the cover of an operator's Instruction Manual. (c.1971)

Remington Electric (Model SR-101). Serial number: N201,392. (c.1974) Availability: Common. Value: $150.

B S I Electric. Manufactured by Business Systems, Inc. (BSI) in Holland. (c.1981). Availability: Scarce. Value: $150.

Remington Rand Electronic (Model 1500). Manufactured by TEC Japan. Availability: Common. Value: $125.

Remington Rand Electronic (Model 1750). Made in Japan. Serial number: 5,780. Availability: Common. Value: $125.

Remington (Model 2000). Manufactured by the Remtronic Division of Remington Industries E Commercio de Sistemec Para Estcretoris South America. Serial number: 0007,001. Availability: Scarce. Value: $200.

Remington Electronic Word Processing System (Model 210 Typewriter & Rem Riter I CRT and floppy disk drive). Manufactured in West Germany for Remington Rand. Availability: Scarce. Value: $200.

It seems appropriate to end this chapter on Remington with a counted cross stitch replica of the Remington Trade Mark. It was created as a gift to the author in July 2001 by Miss Sonya Zawalski to commemorate the completion of his third book about his antique business equipment collection.

Counted Cross Stitch replica of the Remington Trade Mark created by Miss Sonya Zawalski in July, 2001.

Chapter Five

ANTIQUE & COLLECTIBLE TYPEWRITERS OF THE WORLD

Addressograph

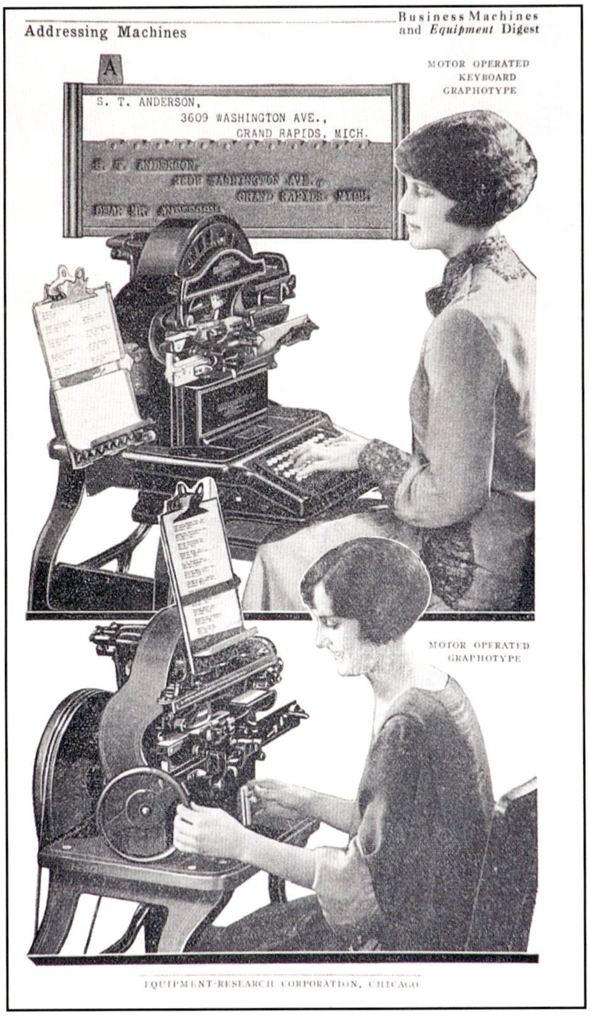

Addressograph (Model - GS-44). Manufactured by the Addressograph Company at Peoria and West Van Buren Street in Chicago, Illinois. (c.1927) A typewriting device that was used to emboss metal plates with names and addresses, which in turn, were used to produce mailing labels at a speed from 1,000 to 10,000 impressions per hour. The above machine was motor-driven and had a standard four-row typewriter keyboard with 44 keys. The alphabetical characters were upper case only and included numerals and punctuation marks. Serial number: 1,289. Availability: Rare. Value: $400.

Adler

Adler (Model 1 of 1912). Manufactured by Klein-Adler Adlerwerke Vorm. Heinrich Kleyer, AG Frankfurt, A/M. Serial number: 245,945. Availability: Common. Value: $250.

Adler (Model 15). Manufactured by Adlerwerke vorm. Heinrich Kleyer A. G. Frankfurt, A/M Germany. Serial number: 282,279. Availability: Scarce. Value: $275.

Adler (Model 7). Manufactured by Adlerwerke Vorm of Frankfurt, Germany. A forward thrust type movement as originally patented by Wellington P. Kidder on March 29, 1892. Patent numbers: 471,794 and 471,795. A universally popular typewriter that was marketed in abundance throughout the world; called the Wellington in the United States, the Empire in Canada, and the Adler in Germany. Serial number: 254,809. (c.1898) Value: $150.

Adler (Model 25). Serial number: 340,644. Value: $225.

Klein-Adler (Model 2 – Portable). Manufactured by Adlerwerke Vorm of Frankfort, Germany. Refer to Adler. (c.1927) Serial number: 362,803. Availability: Scarce. Value: $125.

Adler (Model - 1938). Manufactured by Adlerwerke Vorm of Frankfurt, Germany. Produced with a segment shift, standard type bar action and modern styling. The model pictured above had a six-key decimal tabulator placed directly above the numeral row of keys. Serial number: 583,212. (c.1938) Availability: Common. Value: $75.

Adler (Model - Favorit). Serial number: 336,398. Availability: Common. Value: $100.

Adler (Model - Tippa 1). Manufactured in Western Germany. Serial number: 4567133. Platen length: 9.25", 44-key keyboard. Availability: Scarce. Value: $125.

AEG

A E G (Model 6). Manufactured by AEG Deutsche Werke AG-Berlin of Werk Erfurt, Germany. The Allgemeninen Elextrizitats Gelsellschaft (AEG) company began its existence with the manufacture of the Mignon Index Model Typewriter around 1905. The above Model 6 was produced in the period just before AEG became the Olympia Office Machine Company of Germany. Serial number: 66,905. (c.1930) Availability: Common. Value: $100.

Allen

Allen (Model – A). Manufactured by the Allen Typewriter Company of Allentown, Pennsylvania. This was one of the first machines manufactured by the Allen Typewriter Company with a serial number of 214. This version of the Allen had a straight top plate as opposed to the second model that had a top plate that was cut out to match the shape of the type basket. (c.1918) Availability: Scarce. Value: $200.

Allen (Model - B). Manufactured by the Allen Typewriter Company of Allentown, Pennsylvania. This model also has a three-row keyboard with a curved top plate that distinguished it from an earlier version that had a straight top plate above the type basket. Serial number: 3,859. (c.1918) Availability: Scarce. Value: $200.

Allen, R. C. (Model - Music). Manufactured by the R. C. Allen Business Machines Company an offshoot of the Woodstock Typewriter Company. This particular model was an electric typewriter with a keyboard containing musical notes and symbols. It also had a carbon ribbon mechanism. Serial Number: AC2,233,184. (c.1960) Availability: Scarce. Value: $250.

Alpina

Alpina. Manufactured in West Germany. Platen length: 9.75", 44-key keyboard. Serial number: 201,757. Availability: Common. Value: $50.

American

American (Model - Indicator). Manufactured by the American Typewriter Company at 265 Broadway in New York, New York. In the October, 1923, issue of *Typewriter Topics*, this model of the American is described as an indicator type movement. The indicator was used to select one of the 71 characters that were to be typed. The indicator was located on a semicircular frame with the rubber type directly underneath. By depressing the type key at each selection, a character was imprinted using an ink roll to transfer ink. It was not considered an office or production oriented typewriter. Because of its low purchase price ($10), many were sold and are still found at various outlets today. Serial number: none. (c.1893) Availability: Scarce. Value: $300.

SUNSET—ADVERTISING SECTION

A MACHINE THAT WILL GIVE YOU SATISFACTION FOR YEARS TO COME

See the Point?

THE $100 MACHINE	THE $50 AMERICAN
Universal Keyboard	So Has the American
Print from Ribbon	So Does the American
Type-Bar Machines	So Is the American
Adaptability for Rapid Work	Speed Unlimited on American
Various degrees of Complicated Mechanism	1200 less parts on the American
High Quality of Work	Perfectly Aligned, Clear-Cut Impression from American
Carriage Action—On Some Light, on Others Cumbersome	Extremely Light Ball-Bearing Carriage on American
Portability—Next to Impossible when Convenience is Considered	Fewer parts, Less Weight
Durability—Varied	One-piece Bar, Fewer Parts, Less Complex, Longer Life
Two-Color Ribbon Shift—Only on some machines	Two-color Ribbon Shift on every American
Tabulating Indicator—Some have it, Some Charge Extra	Tabulating Indicator on every American
PRICE, $100	OUR PRICE, $35

OUR SPECIAL PRICE, $35, SAVES YOU $65

In simplicity of construction the American stands alone.

Light and compact, portable in the true sense of the word.

Equally adapted to the office, study or home.

Productive of the best work; in itself pleasing in design and finish and of the strongest construction.

Capable of the greatest speed and learnable in a few hours of practice without an instructor.

The only direct acting type-bar machine having key and type on one solid steel bar.

Furnished in all languages, French, German, Spanish, Russian, Greek, etc.

Ink ribbon instantly changeable without soiling the hands. Will write in any color required.

Equipped with ball-bearings for light running.

Usable anywhere without a special table. Special Club Price, $35.

Model No. 8

REGULAR PRICE $50 SPECIAL PRICE $35

A Compact, Efficient, Durable Typewriter—The result of years of typewriter-making experience

THE AMERICAN TYPEWRITER contains a number of exclusive features that make it in **Simplicity, Durability and Efficiency**, all that could be desired.

In appearance **The American** is a beauty. All exposed steel parts are highly nickeled while other portions of the machine are heavily enameled. Either white keys with black letters or black keys with white letters are furnished as desired. Rubber dust covers provided with all machines.

Sunset August,

American Typewriter Club,
William Peck Co.,
2803 Folsom St., S.F.

Gentlemen: I accept the special offer to Sunset readers of a regular $50 American Typewriter for $35. Send me one of these typewriters f. o. b., San Francisco. It is understood that I may try the machine for five days. If satisfactory, I agree to remit $5 within five days of receipt, and pay the balance price in SIX monthly installments of $5 each. If the typewriter is unsatisfactory, I agree to repack carefully and notify within five days, holding subject to your order. Title of the American Typewriter to remain in your name until the machine is fully paid for.

Name ....

Address ....

Town .... *State* ....

References ....

IN WRITING TO ADVERTISERS PLEASE MENTION SUNSET

American (Model 7). Manufactured by the American Typewriter Company at 265 Broadway in New York, New York and in the Williams Plant in Derby, Connecticut. "Keep It Simple" was the mission of N. R. Marshman and Lee Burridge when inventing the type bar version of the American Typewriter. Simple it was, as they reduced the number of parts used to well under 350 and in so doing allowed for a much smaller footprint. The American was less than half the height of the conventional standard typewriter of the period and weighed at least half of one of the leading sellers of the day, Remington. One of the unique mechanisms, or we should say lack of mechanisms, was the American type bar. It was a double shift type with a type bar, a key top, and type, all in one piece which effectively eliminated all the parts and linkage that were normally found on the type bar mechanism. By doing so, the manufacturer may have doomed their chance of success because they created a different touch for operators. As a result, the American Typewriter Company closed their doors in 1915 after only 14 years in business. Serial number: 17,033. (c.1899) Availability: Scarce. Value: $400.

American (Model 8). Manufactured by the American Typewriter Company at 265 Broadway in New York, New York. In 1908, after initially producing typewriters at the Williams factory in Derby, Connecticut, the American Typewriter Company moved their manufacturing facility to Bridgeport, Connecticut. At the same time, they announced the release of the Model 8. Serial number: 23,997. (c.1908) Availability: Scarce. Value: $400.

American Flyer

American Flyer. A toy typewriter manufactured by American Flyer Manufacturing Company of Chicago, Illinois. This is the same company that made model trains. (c.1935) Availability: Common. Value: $25.

American Visible

American Visible (Model - Index). Manufactured by the American Typewriter Company at 265 Broadway in New York, New York. The American Visible has an inclined metal front plate with 46 keys permanently displayed in the standard keyboard arrangement. A nickel-plated selector device, especially designed to fit the curve of the operator's fingertip, could be moved to select the character desired and depressed to type that character. Felt pads were used for inking. It was also manufactured under the name Young American. (c.1891) Availability: Rare. Value: $1,000.

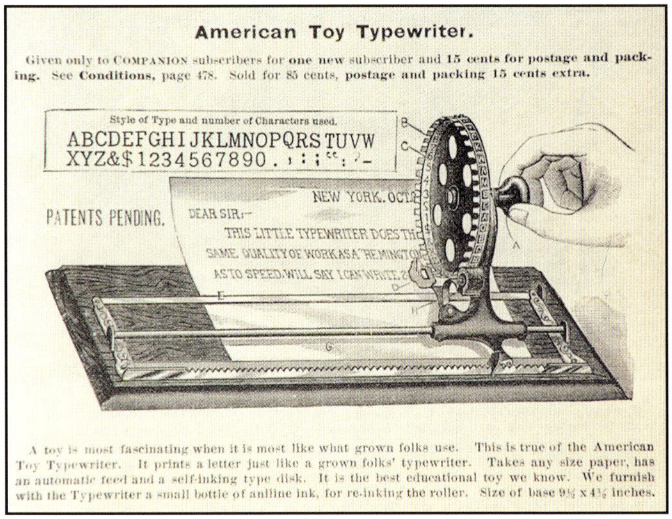

Advertisement from the *American Boy* magazine of November 1915.

Anderson, G.K.

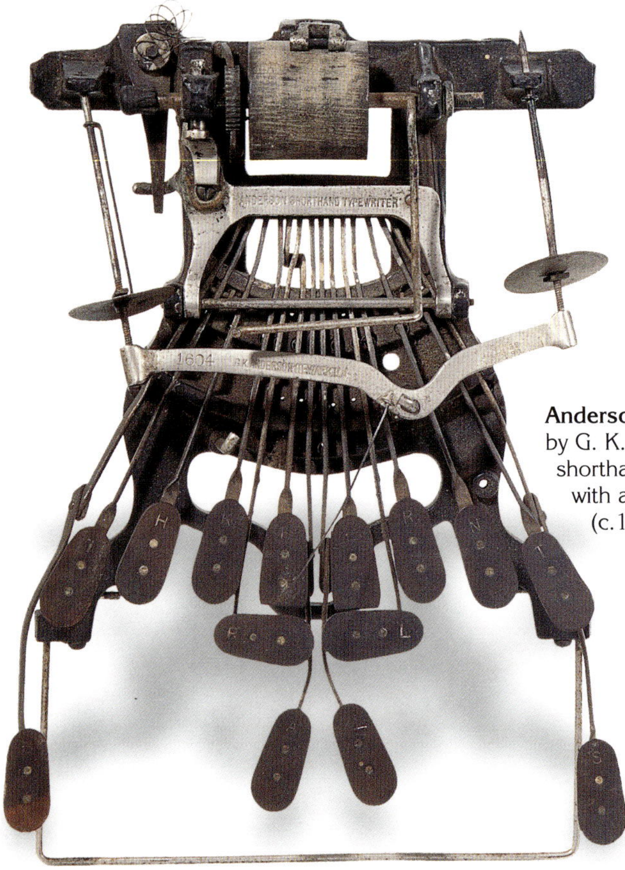

Anderson, G. K. (Model - Shorthand). Manufactured by G. K. Anderson. A very small typewriter incorporating shorthand symbols, using 14 keys and a paper tape with a cloth ribbon for inking. Serial number: 1,604. (c.1889) Availability: Rare. Value: $1,000.

Annell'

Annell' (Model 3A). Manufactured by the Woodstock Typewriter Company of Woodstock, Illinois. Reported to be a copy of the Woodstock Model 4 typewriter that was marketed by Harry A. Smith through mail order channels. Serial number: 1,789. (c.1922) Availability: Scarce. Value: $300.

Automatic Printing Wheel

Automatic Printing Wheel. Manufactured by Stenhouse & Company of Chicago, Illinois. This was a device for printing on cartons. Availability: Scarce. Value: $50.

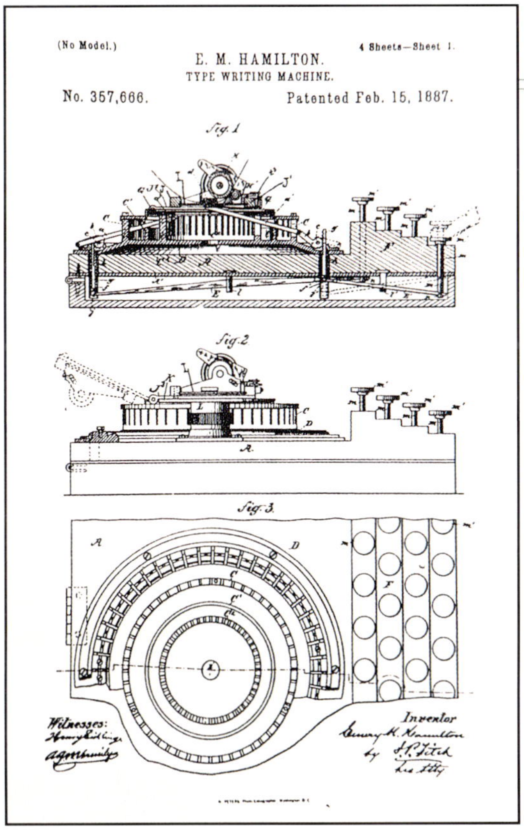

Automatic

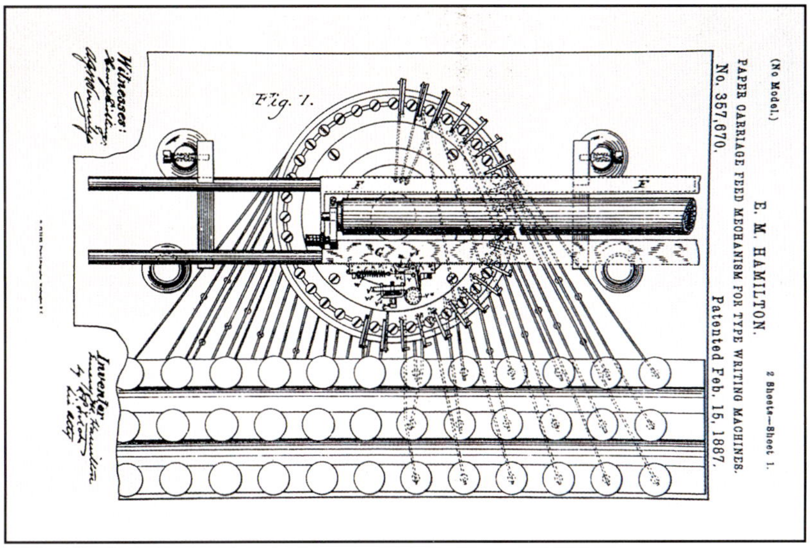

Automatic. Manufactured by the Hamilton Typewriter Company of New York, New York. Invented by Emery M. Hamilton of New York City and patented on February 15, 1887. The Automatic was a small brass machine that weighed only 11 pounds and measured 11" x 8" x 14." It has a standard universal keyboard with 3 rows and 48 keys. An upstroke machine with type bars that measure only 1 ½ " long. It typed in proportional or differential spacing utilizing upper case only. A unique feature was that the space bar was directly behind the top row of keys and displayed the name of the machine. Patent number: 357,666 dated February 15, 1887. The automatic is a wonderful specimen of early American mechanical artistry, similar to that found in old precision brass clocks. Really must be seen to be appreciated; unfortunately, few are known to have survived. It is considered one of the most desirable and collectible typewriters. Serial number: 168. (c.1881) Availability: Very Rare. Value: $15,000+.

Autotypist

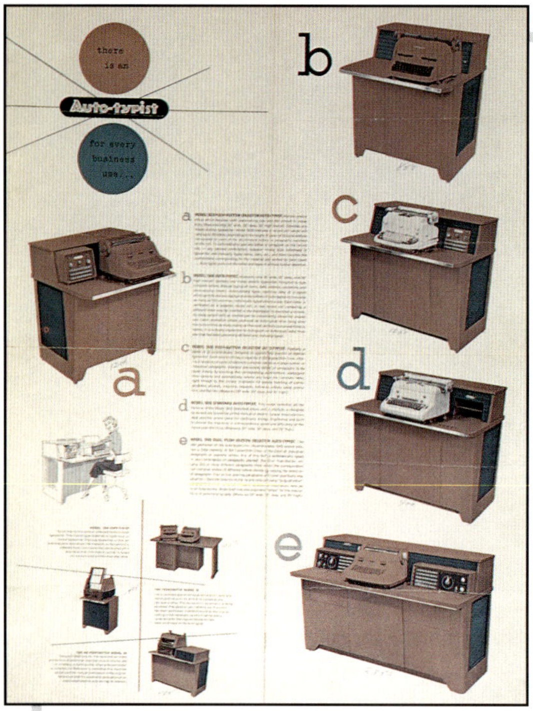

Autotypist. Manufactured by the American Automatic Typewriter Company of Chicago, Illinois. In 1932, Otto Schultz, a player piano manufacturer attended a business show and saw a demonstration of the pneumatic action [used in player pianos] attached to a typewriter. The company that developed this product suffered in the 1929 financial collapse and had gone through bankruptcy. Schultz purchased the rights to the product and moved its manufacturing to his piano factory in Chicago, Illinois. A few years later, Schultz persuaded a teletype company engineer, Henry Kloid, to join him as head of Research and Development (R&D). By 1934, they had created the Autotypist. An automatic typewriter with a selector dial. The operator could then select any one of ten (10) different form letters and personalize it with the insertion of an individual's name and address. That same year, an electric typewriter was attached to an Autotypist. Through the years, the Autotypist was refined to the point that it could store 50,000 characters, had a dual roll and a 100 push button selector panel. The main application was form letters. Individual paragraphs were stored on tape. These paragraphs were then retrieved and used to complete individualized letters. The Autotypist Company had effectively positioned itself as a pioneer in the new "word processing" market. Serial number: 019. (c.1950) Availability: Rare. Value: $500.

Bar-Let

Bar-Let (Model 2). Manufactured by the Bar-Lock Company of Nottingham, England. (c.1925) Availability: Common. Value: $150.

Bar-Lock

Bar-Lock (Model 4). Manufactured by the Columbia Typewriter Company of New York, New York. Patented by Charles Spiro in 1889, the Bar-Lock series had a beautifully sculptured front shield and was one of the most artistic typewriters ever produced. It derived its name from a feature that boasted to provide perfect type alignment. It consisted of a series of pins that were located at the center of impact for each type bar and effectively forced the type bar between the pins, therefore producing perfect alignment. The Bar-Lock has a double keyboard and the type bars stand straight up in a semicircle. They strike forward and down toward the platen. Serial number: 10,281. (c.1895) Availability: Rare. Value: $2,000.

Bar-Lock (Model 6). Manufactured by the Columbia Typewriter Company of New York, New York. For a description please refer to the Bar-Lock Model 4. Serial number: 16,313. (c.1895) Availability: Rare. Value: $2,000.

Barr Universal

Barr Universal. Manufactured by the Barr-Morse Corporation of Ithaca, New York. Invented by John M. Barr and first manufactured in New York and later in Canada. A four-bank keyboard typewriter that was light enough to be a portable. Serial number: unknown. (c.1926) Availability: Scarce. Value: $125.

Bennett

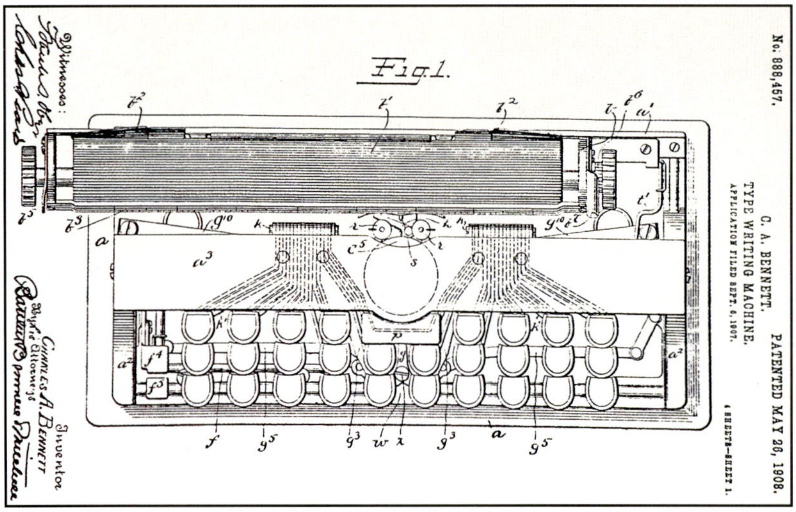

Bennett (Model - Black). Manufactured by the Bennett Typewriter Company of Harrisburg, Pennsylvania. Invented by Charles Bennett of Dover, New Jersey. By today's standards, the Bennett would be called a Notebook Portable. It was very small and compact with a leather carrying case. It was originally introduced as the Junior typewriter and was inked by two rollers. The Bennett model was patented in 1907 with a ribbon inking system. Serial number: 20,576. (c.1910) Availability: Common. Value: $300.

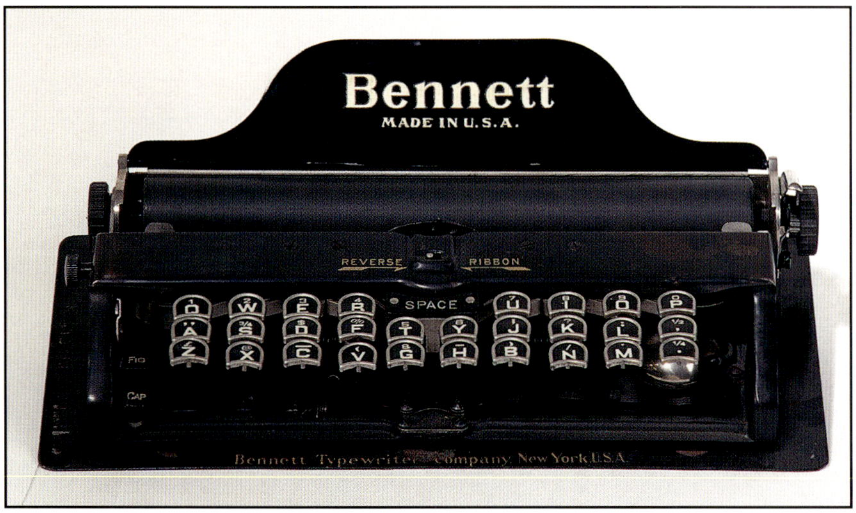

Bennett (Model - Aluminum). Manufactured by the Bennett Typewriter Company of Harrisburg, Pennsylvania. Refer to Bennett Black Model for description. Serial number: 13,435. (c.1910) Availability: Common. Value: $300.

Bijou

Bijou (Model - Folding). Manufactured by Siedel & Naumann, A.G.A portable with a three-bank keyboard. The Standard Folding Typewriter Company had previously introduced the folding feature. Serial number: 82,847. (c.1915) Availability: Common. Value: $150.

Bing

Bing (Model 1). Manufactured by Bingwerke, A. G. of Germany. A compact portable fabricated in a manner similar to many of the toy products of the period. The Model 1 used a felt roller to ink the type. The Model 1 is more difficult to find than the Model 2 that used a ribbon to distribute the ink. The Model 2 shows up frequently at flea markets as well as other marketing channels. Serial Number: Unknown. (c.1920) Availability: Rare. Value: $300.

Bing (Model 2). Manufactured by Bingwerke, A. G. of Germany. The Model 2 used a fabric ribbon instead of the ink roller used on the Model 1. Refer to Bing Model 1 for more information. (c.1920) Serial number: Unknown. Availability: Common. Value: $50.

Blickensderfer

George C. Blickensderfer was one of the prominent typewriter inventors of the period as well as one of the most successful, with a career that spanned 28 years (1889 to 1917). His success was built around the principle of substituting a type ball or wheel with 84 engraved characters and used an ink roll to print on paper. This was not a new principle but Blickensderfer's invention was so well done it seemed new. It had 28 keys and used a double shift mechanism. It was portable. Its durability is evident in the fact that it is still found in abundance everywhere. It is unfortunate that many collectors consider many models of the Blickensderfer as common; but on the other hand it is wonderful that almost every collector has one or more in his collection. Among the models known to have been produced are: Models 5, 6, 7, 8 and 9. Manufactured in 1893, 1896, 1897, 1907 and 1914 respectively. A music typewriter, the "Naco Blick" appeared in 1910. An oriental and electric were also released. A type bar machine was released in 1916, the Blick Bar. In 1919 a small portable, the Blick 90 was released. It was also produced under the name Roberts 90, named after its inventor.

Blickensderfer (Model 5 and Model Visible). Manufactured by the Blickensderfer Manufacturing Company of Stamford, Connecticut. Invented by George C. Blickensderfer of Erie, Pennsylvania. Serial number Model 5: 8,616. (c.1893) Availability: Common. Value: $150. Serial number Blickensderfer visible: 163,818.

Blickensderfer (Model 6 - Aluminum). Manufactured by the Blickensderfer Manufacturing Company of Stamford, Connecticut. This aluminum model was designed to minimize its weight and to make the typewriter more marketable from a portable standpoint. See Blickensderfer Model 5 for more details. Serial number: 158,153. (c.1896) Availability: Common. Value: $175.

Blickensderfer (Model 7 – Decorated). Manufactured by the Blickensderfer Manufacturing Company of Stamford, Connecticut. This model is a little unusual. It had an elaborately stencil-decorated case. Serial number: Unknown. (c.1897) Availability: Scarce. Value: $250.

Blickensderfer (Model 7 – Black). Manufactured by the Blickensderfer Manufacturing Company of Stamford, Connecticut. The Model 7 was popular and available in several different varieties. The machines pictured are in the author's collection and are representative of some of the different model 7s manufactured. Serial number: 74,845. (c.1897) Availability: Common. Value: $125.

Blickensderfer (Model 7 – Aluminum). Manufactured by the Blickensderfer Manufacturing Company of Stamford, Connecticut. Serial number: 145,645. (c.1897) Availability: Scarce. Value: $250.

Blickensderfer (Model 7 – Wide Carriage). Manufactured by the Blickensderfer Manufacturing Company of Stamford, Connecticut. Serial number: 101,089. (c.1897) Availability: Scarce. Value: $200.

Blickensderfer (Model 9). Manufactured by the Blickensderfer Manufacturing Company of Stamford, Connecticut. The Model 9 was only manufactured for one year and therefore is very difficult to find. Serial number: 184,581. (c.1918) Availability: Scarce. Value: $250.

Blickensderfer (Model 8 – Decimal Tab). Manufactured by the Blickensderfer Manufacturing Company of Stamford, Connecticut. This model has a decimal tabulator just to the left of the typing element. Serial number: 136,518. (c.1908) Availability: Common. Value: $175.

Blickensderfer (Model - Featherweight Aluminum). Manufactured by the Blickensderfer Manufacturing Company of Stamford, Connecticut. A light-weight, aluminum model appropriately named the Featherweight . Serial number: 176,514. (c.1910) Availability: Common. Value: $200.

Blickensderfer (Model - Home). Manufactured by the Blickensderfer Manufacturing Company of Stamford, Connecticut. The Home Blick was similar to the Model 5 without many features. It was sold through the Montgomery Ward Department Store. Serial number: 177,743. (c.1917) Availability: Common. Value: $200.

Blickensderfer (Model - Oriental). Manufactured by the Blickensderfer Manufacturing Company of Stamford, Connecticut. The Oriental is a rare version of the Blickensderfer family of typewriters and as a result is considered very collectible. Patent date: April 12, 1892. Serial number: 472,693. (c.1897) Availability: Rare. Value: $1,000.

Blick-Bar. Manufactured by the Blickensderfer Manufacturing Company of Stamford, Connecticut. The Blick-Bar was the invention of Emmett G. Latta whose patents were assigned to the Blickensderfer Manufacturing Company. This was the company's first type-bar machine. The company began manufacture in 1913 but production was delayed because of the start of World War I. By 1916, production began to increase. The Blick-Bar was off to a good start, but in 1917 George C. Blickensderfer died. The company could not survive without its founder and business ended shortly thereafter. It was only in production for about one year before it was picked up by Harry A. Smith and marketed through his mail order channels for a brief period. (c.1916) Availability: Rare. Value: $400.

Blick 90. Introduced in 1919 and named for the 90-character keyboard. Serial number: B-547. Availability: Rare. Value: $1,700.

Blista

Blista (Model - Braille). Manufactured in Blindenstudienstalt, Marburg, Lahn – Germany. Serial number: 3,878. (c.1950) Availability: Common. Value: $125.

Brooks

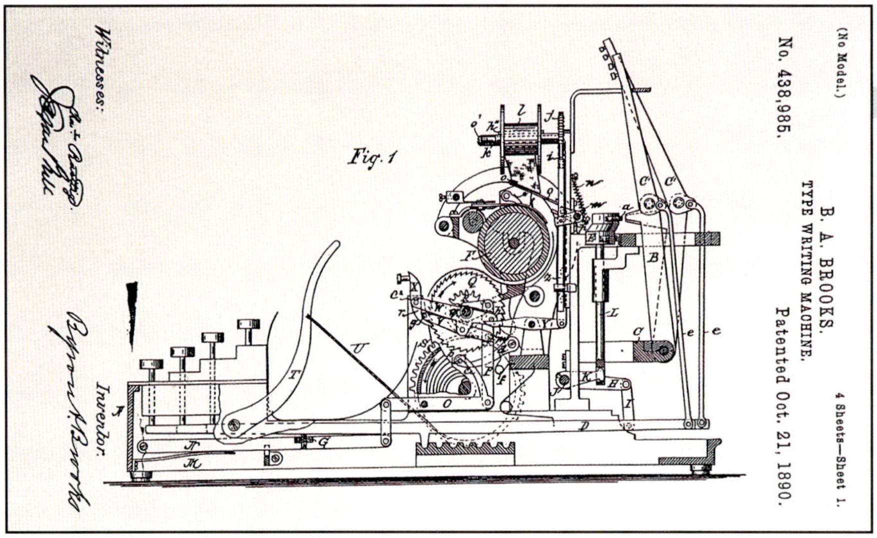

Brooks. Manufactured by the Union Typewriter and Supplies Company. The Brooks was an interesting typewriter. Its appearance was substantially different from the conventional typewriters of the period. Type bars, each with three characters, were arranged in a semicircle and located directly behind and above the platen. The rear down stroke action was unique but not exclusive since it also could be found on the Fitch and Waverly. The main marketing thrust of the Brooks was based on its light touch, light weight (less than 18 pounds) and visible writing line. Byron Alden Brooks of Kings County, Brooklyn, New York, was the inventor. His first patent was granted on August 4, 1885. Patent number: 323,495. Brooks is credited as one of the two inventors of Remington's first shift mechanism. It was Brooks's idea to mount two (2) characters (upper & lower case) on the same type bar. Brooks's 1885 patent was for the first typewriter successfully marketed using the down stroke principle. The type and bars were arranged in a semicircle behind and above the platen. The type was visible to the operator without lifting the platen. There were four Brooks patents granted: 323,495, August 4, 1885; 423,900, March 25, 1890; 438,985, October 21, 1890; and 454,845, June 30, 1891. Serial number: 911. (c.1887) Availability: Rare. Value: $9,000.

Brosette-Portable

Brosette Portable. Manufactured in Germany. Platen length: 9.1875", 43-key keyboard. Serial number: 20,851. Availability: Common Value: $75.

Burnett

Burnett. Originally introduced as the Triumph Visible in 1907 and manufactured by the Visible Typewriter Manufacturing Company of Kenosha, Wisconsin. The company went into receivership the same year. It was then manufactured in 1907 by the Burnett Typewriter Company of Chicago, Illinois and marketed through Sears, Roebuck and Company. (c.1908) Serial number: 990. Availability: Rare. Value: $12,000.

Burroughs

Burroughs (Model 50A). Manufactured by the Burroughs Adding Machine Company. They had a successful career in manufacturing adding machines and calculators. Eventually Burroughs propelled itself into the typewriter market by acquiring the Moon-Hopkins combination adding machine and typewriter. They probably felt they could diversify and be more competitive by introducing a typewriter. The manual typewriter, introduced in 1931, was not particularly innovative. Although a respectable number of machines were sold (probably because of Burroughs strong marketing base) it was discontinued and Burroughs concentrated on their adding machine and calculator business. Serial number: 129,623. (c.1931) Availability: Common. Value: $50.

Burroughs Electric (Model - A). An electric version (carriage only) was introduced in 1932. See Burroughs Model 50A for more details. Serial number: 350,491. (c.1932) Availability: Common. Value: $125.

Butler

Butler. Manufactured by the Lambert Typewriter Company of Brooklyn, New York. See Lambert. Serial number: 6,976. (c.1895) Availability: Scarce. Value: $1,000.

Caligraph

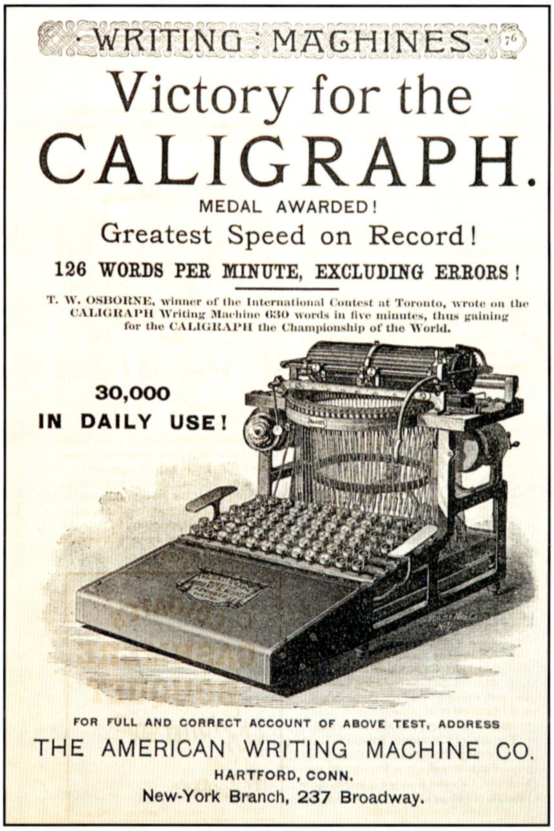

Caligraph (Model 2). This version was larger than the Model 1. It also had a 72-character keyboard. Serial number: 877. Availability: Scarce. Value: $200.

Caligraph (Model 1). Manufactured by the American Writing Machine Company of New York, New York. The invention of Franz C. Wagner and George Washington Newton (sometimes called Napoleon) Yost, it was the second typewriter to appear on the American market. It was a down-stroke machine with a circular type basket similar to the Sholes & Glidden as well as the Remington Blindwriter models. It was also boasted to be the first machine to have a separate key for every character. It was a lightweight machine with a light touch and competed favorably in many of the initial speed typing contests against the Remington. Serial number: 2,905. (c.1882) Availability: Rare. Value: $5,000.

Caligraph (Model 4). Manufactured by the American Typewriter Company of Bridgeport, Connecticut. This was a larger version of the Model 1 Caligraph. Serial number: 9,843. (c.1885) Availability: Scarce. Value: $300.

Cardinal

Cardinal. Manufactured in 1921 by the watch and clock factory, L. Furtwänger & Sons, Baden, Germany. The machine was 42 keys, ball bearing type and 9.5" platen. Serial number: 0631. Availability: Scarce. Value: $850.

Century

Century (Model 10). Manufactured by the American Writing Machine Company of New York and actually built in the Smith Premier Factory in Syracuse, New York. A small portable double shift machine that weighed less than 18 pounds. Serial number: HM644. (c.1919) Availability: Common. Value: $100.

Cash

Cash. *See Typograph.*

Champion

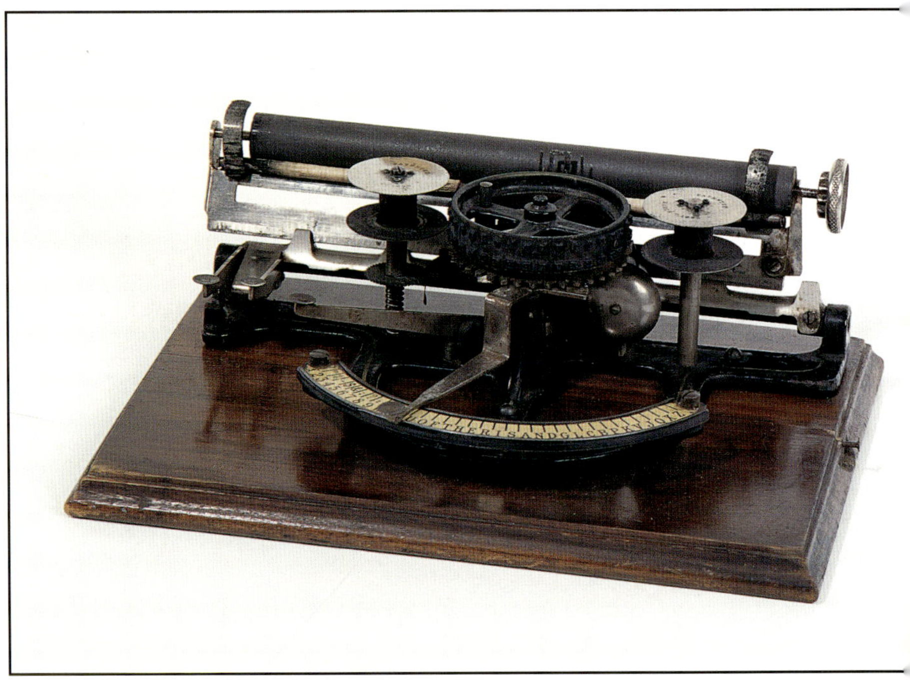

Champion. Manufactured by the Garvin Machine Company of New York, New York. A later version of the Peoples typewriter with a different method of handling the ribbon. The Champion utilized a regular ribbon for inking and the Peoples used an inked roll. (c.1893) Availability: Rare. Value: $1,000.

Chicago

Chicago (Model 3). Manufactured by the Chicago Writing Machine Company at 85 Wendell Street in Chicago, Illinois. Serial number: 1,903. (c.1900) Availability: Rare. Value: $3,000.

Chicago (Model 1 – Decorated). Manufactured by the Chicago Writing Machine Company at 85 Wendell Street in Chicago, Illinois. Stencil decorations are painted around the face plate name. (c.1902) Availability: Scarce. Value: $900.

Chicago (Model 1 – Pin Stripe). Manufactured by the Chicago Writing Machine Company at 85 Wendell Street in Chicago, Illinois. Serial number: 72,35. (c.1900) Availability: Scarce. Value: $800.

Coffman

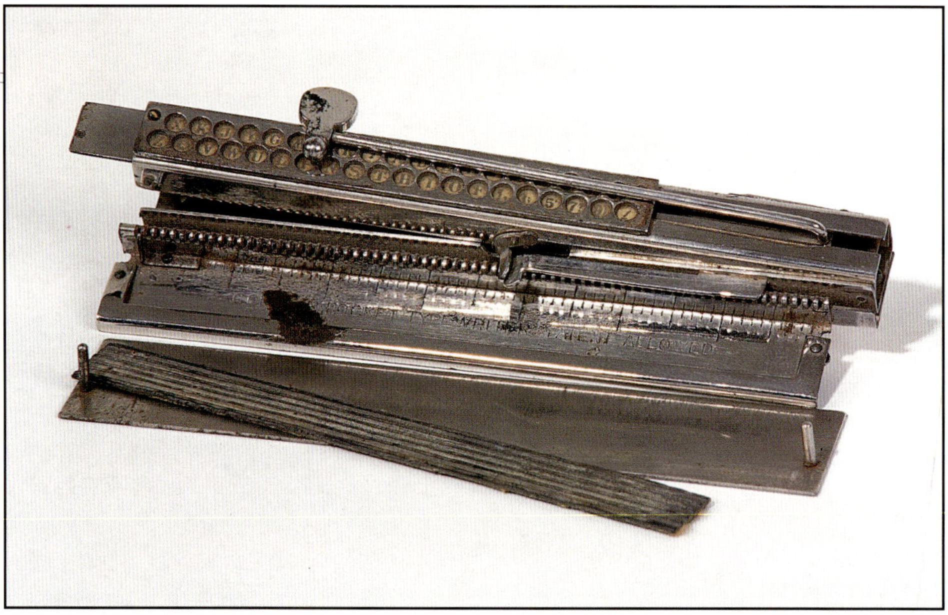

Coffman (Model (a) – without Platen). Manufactured by the Coffman Manufacturing Company at 529 Spruce Street in St. Louis, Missouri. G.W. Coffman of Garden City, Kansas patented the machine in 1902. Weighing only 20 ounces and measuring less than 10 inches, the Coffman could be carried in your pocket and marketing emphasized that virtue. There were two versions of the Coffman: one with a platen and one without. Serial number: 000. (c.1902) Availability: Rare. Value: $4,000.

Columbia

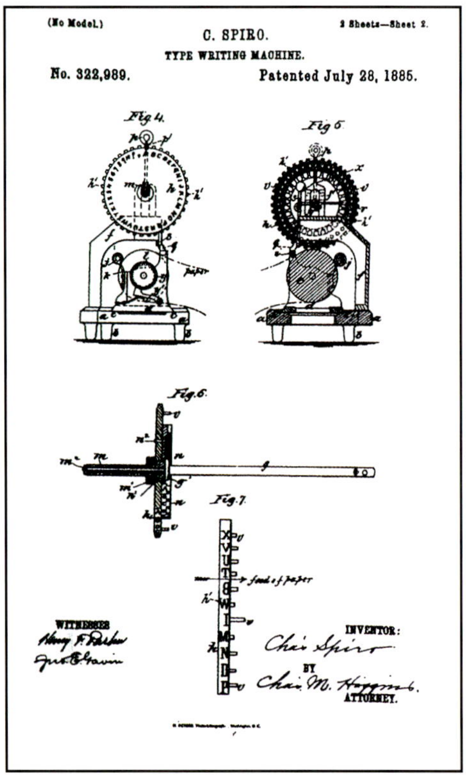

Columbia Index (Model 2). Manufactured by the Columbia Typewriter Company at 129-131 Crosby Street in New York, New York. Charles Spiro patented the Columbia Index machine on February 2, 1886. Patent number: 335,392. He was also the designer of the Bar-Lock and other typewriters. Two models of the Columbia are known to exist: one with a single (all caps) printwheel and one with a double (upper & lower case) printwheel. Serial number: 1,127. (c.1886) Availability: Rare. Value: $4,000.

Columbia Bar-Lock (Model 8). Manufactured by the Columbia Typewriter Company of New York, New York. Serial number: 47,547. Availability: Scarce. Value: $300.

Columbia Bar-Lock (Model 10). Manufactured by the Columbia Typewriter Company at 116th Street at Fifth and Lenox Avenue, New York, New York. Another example of Charles Spiro's inventions. He was issued patents in 1891 and 1892. Also see Royal Bar-Lock. Serial number: 90,885. (c.1905) Availability: Scarce. Value: $300.

Commercial

Commercial Visible (Model 6). Manufactured by the Commercial Visible Typewriter Company of New York, New York. Invented by R. W. Uhlig, the Commercial Visible used a type-wheel. The result was clearly visible. It had 28 keys, a double shift and a universal keyboard. Serial number: 22,132. (c.1898) Availability: Rare. Value: $2,500.

Conover

Conover. Manufactured by Lowar, Spencer, Bartlett & Company of Chicago, Illinois. The Conover typewriter was originally marketed as the Chicago. Additional names under which it was marketed are: Ohio, Galesburg, Yale, Draper and Baltimore. Serial number: 5,776. (c.1900) Availability: Rare. Value: $800.

Continental

CONTINENTAL-RAPIDus-Schreibmaschine mit Zählwerken

Continental. Manufactured by the Wanderer Fahrradwerke of Schonau, Germany. The Wanderer Company began selling motorcycles and milling machines in 1894. They introduced their typewriter in 1904. By 1913 they had sold over 60,000 units. Eventually their product line included the Silenta, the Standard, and the Portable (all shown above). Serial numbers: Silenta - 752,308; Standard – 974,920; Portable – 263,975. (c.1904) Availability: Common. Value each: $50-$150.

Corona

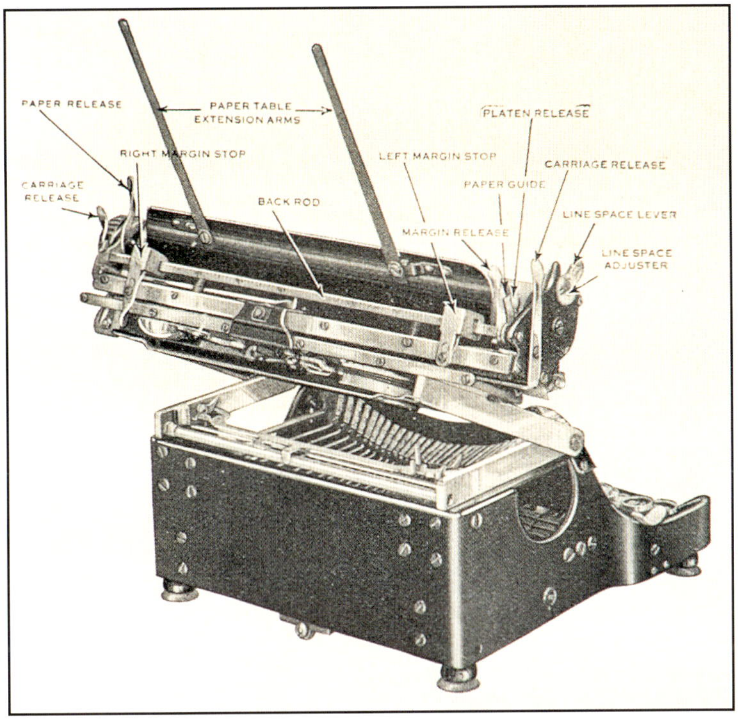

Corona (Model 3). Manufactured by the Corona Typewriter Company of Groton, New York. Originally introduced in 1906 as the Standard Folding Typewriter. A few years later reappeared as the Corona. Because of the popularity of this typewriter and the large number manufactured, it is one of the most common machines found in antique malls and flea markets. Serial number: 119,897. (c.1912) Availability: Common. Value: $50.

CORONA
in six
beautiful Duco colors
to suit your taste
If atmosphere has anything to do with the way you think and write—this gay Corona portable will simply breathe life into the words you type.
Corona can now be in perfect harmony with your library, your boudoir, or the little nook of a study where you write. You may choose from a variety of exquisite color combinations—devised to suit people, like yourself, who love beauty in everything.
Mechanically these new colored Coronas are as nearly perfect as typewriters can be, embodying as they do, all of our twenty years of experience in this field.
Seeing is believing. Step into the nearest Corona store and feast your eyes on these beautiful new models.
These are the six colors in which Duco finished Coronas are obtainable:
1 Lavender. Panelled in crystalline rose gold.
2 Light Maroon. Crystalline panels in same color.
3 Channel Blue. Crystalline panels in same color.
4 Mountain Ash Scarlet. Contrasting panels in crackle-finish black. This is the finish shown in the large picture above.
5 Cream Color. Panelled in crystalline rose gold.
6 Bruce Green. Crystalline panels in the same color.
Write for our illustrated folder showing all six of these new Duco finished models in full, rich colors. Better still, look for Corona in your phone book and ask to have a colored Corona brought to you for examination. Used machines will be accepted in trade and easy terms arranged to suit you.
L C Smith & Corona Typewriters Inc
Sales offices in principal cities of the world
1002 E Washington Street Syracuse N Y

CORONA
SPECIAL
L C Smith & Corona Typewriters Inc

CORONA
SPECIAL
L C Smith & Corona Typewriters Inc

Corona (Special Models). Manufactured by the L. C. Smith & Corona Typewriter, Inc. Serial numbers: Blue – BX 664,930; Gold – GX 651,543; Green – BX 623,046 and Red – SX 618,664. (c.1920) Availability: Common. Value each: $150.

Corona (Model 4). Manufactured by L. C. Smith & Corona Typewriter Inc. Released for the 1930 Christmas season. It was available in six different colors in addition to black. Serial number: Gold – 6,001; Red – C2L10,979. (c.1930) Availability: Common. Value each: $150.

THE SATURDAY EVENING POST

A Gift "to help little fingers guide eager young minds"

The "Animal Keyboard" is the latest (as the original "folding portable" of twenty-five years ago was the first) in a long line of innovations conceived by Corona. Today's Corona models, offering a wide choice in price and equipment, are briefly described below; each measures up fully to the most exacting Corona standards.

CORONA SILENT.* Thanks to scientific silencing, superb workmanship—our finest portable. Long list of desirable features. $67.50

CORONA STERLING.* Our most highly-perfected non-silent model. Offered, like the Silent, in three optional colors. $62.50

CORONA STANDARD.* A new model, rapidly acquiring fame as "the buy of the year." Has many features found in other higher-priced machines. $49.50

* The three Corona models above have the famous Floating Shift and Touch Selector. In fact, only Corona offers these features in models at or below $50

CORONA JUNIOR. The perfect "in-between" portable for those who want a 4-row Keyboard at lowest cost. $37.50

CORONA THREE. The history-making Corona! First successful portable. 600,000 sold! 3-row Keyboard—but *complete*, and light. $32.50

(All prices subject to revision)

1. *Nine colored finger-rings, adjustable in size.*
2. *The "Animal Keyboard" (applied to a Corona Silent).*
3. *Play-exercise book, prepared by educators.*
4. *Descriptive booklet, free to parents or teachers.*

The new "Animal Keyboard," applied to regular Corona models at nominal extra cost (which also includes finger-rings and play-exercise book). Makes play of correct touch-typing and the first steps in education . . . and yet adults can use it just as they use any other Corona.

This extraordinary development—the Corona "Animal Keyboard"—makes correct touch-typing almost automatic, from the kindergarten up, through any adult age. But much more important—for the first time, we think, it makes the typewriter a practical aid in education.

Already we know it is irresistible to children—bright colors, mechanical movement, animal keys, finger-rings. And see how simply it works!

What finger should write "g"? The Bear finger, of course, wearing the violet ring; instantly the left forefinger finds the correct key. And without drudgery—almost without conscious effort—fingers learn to fly!

Children simply love to type—and we know from a three-year research by educators that ability to type opens the door to quicker learning and largely increased love of writing.

A child could have no gift more fascinating—and perhaps none of greater lifelong value! True for adults, too!

. . .

A limited number of "Animal Keyboard" Coronas in three models will be obtainable from Corona dealers before Christmas; larger production shortly. Liberal terms, as on all Corona models.

Parents and teachers of primary children will find much of interest in the descriptive booklet illustrated. Just check coupon for free copy.

L C SMITH & CORONA TYPEWRITERS INC
Desk 11, 703 E. Washington St., Syracuse, New York
Please send free booklet describing ☐ Animal Keyboard Corona ☐ other Corona models
Name
Address
City State

Corona Animal Keyboard Rings. Manufactured by L. C. Smith & Corona Typewriter, Inc. as a "fun way" to help young typewriter users select the proper finger to depress a key on the keyboard. Each ring represented an animal and corresponded to the same animal on the keyboard. There were only nine (9) rings since one thumb remained ring-less. Modeling the nine (9) rings is the author's granddaughter, Michelle Phillips. Availability: Rare (only set known to exist). Value: $1,000+.

Corona Animal Keyboard (Model - Silent). Manufactured by L. C. Smith & Corona Typewriter, Inc. The above advertising was from the Saturday Evening Post, November 30, 1935. Introduced in time for the 1935 Christmas Season. This portable was unusual in that it had an animal keyboard and came with nine (9) finger rings with animals on them to match the keyboard (see Corona Animal Keyboard Rings below). Designed for kindergarten children and up. Availability: Rare. Value: $600.

Corona Junior Portable. Manufactured by L. C. Smith & Corona Co. Serial number: IJ2037. Platen: 9.75", 42-key keyboard. There is no tab, backspace or margin release key. Availability: Scarce. Value: $250.

Crandall

Crandall (Model - Early Syracuse). Manufactured by the Crandall Typewriter Company of Syracuse, New York. The invention of Lucian Stephan Crandall, it was first patented in 1879 and subsequently improved. The model shown above was the first of the improved models. It was decorated with stenciling and had square aluminum key tops. The Crandall was considered by many to be the first single element typewriter. However, instead of a type ball, it used a rectangular sleeve. The specimen pictured is one of the lowest serial numbers known to exist. Serial number: 1,027. (c.1895) Note: The pictured machine has been completely disassembled, restored, and reassembled. Serial number: 1207. Availability: Rare. Value: $5,500.

Crandall (Model - New). Manufactured by the Crandall Typewriter Company of Syracuse, New York. This Crandall is considered to be the most attractive model Crandall. It was beautifully decorated with gold stenciling and mother-of-pearl inlay. Serial number: 5,562. (c.1895) Availability: Rare. Value: $5,000.

Dactyle

Dactyle. Manufactured by the Dactyle Manufacturing Company at 46 Boul d Haussman in Paris, France and is basically the French version of the Blickensderfer. Serial number: 1,615. (c.1893) Availability: Scarce. Value: $350.

Daugherty

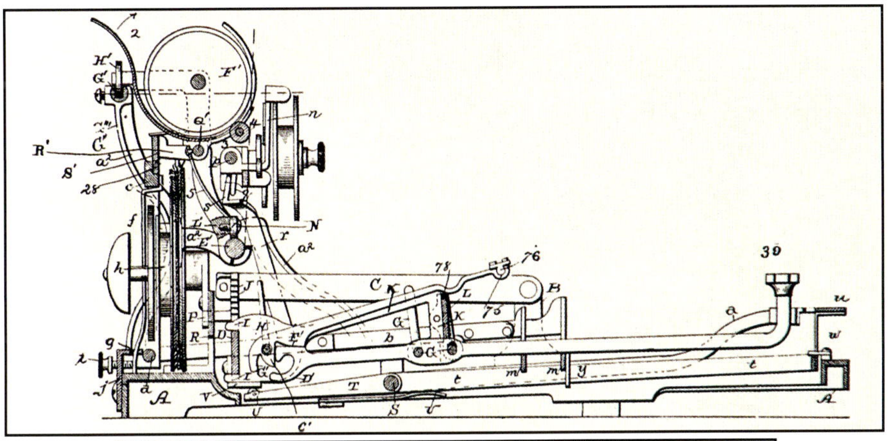

Daugherty Visible. Manufactured by the Daugherty Typewriter Company of Kittaning, Pennsylvania. Invented by James D. Daugherty who was a stenographer. A single shift machine with visible typing. Serial number: 3,148. (c.1890) Availability: Scarce. Value: $750.

Demountable

Demountable (Model 2). Manufactured by the Demountable Typewriter Company of Fond du Lac, Wisconsin. The Demountable was actually developed by the Rex Typewriter Company. Rex had purchased the National Portable Typewriter Company in 1916. In 1918, they started work on the Demountable. Finally, in 1921, they released their first models. The Demountable was a remarkable design. With little effort, it could be separated into three (3) basic units for easy cleaning and maintenance. Serial number: 125,958. (c.1921) Availability: Scarce. Value: $200.

Densmore

Densmore (Model 1). Manufactured by the Densmore Typewriter Company of New York, New York. Introduced in 1891, the Densmore was the invention of Walter J. Baron. It was an under stroke machine or blindwriter. It had 38 keys and printed 76 characters. Because of a carriage that glided on ball bearings, the carriage movement was exceptionally smooth and easy to move back and forth. Serial number: 1,160. (c.1891) Availability: Rare. Value: $1,000.

Densmore (Model 2). Manufactured by the Densmore Typewriter Company of New York, New York. The Model 2 had 42 keys and 84 characters and was manufactured for the European market. Serial number: 8,838. (c.1897) Availability: Scarce. Value: $500.

Densmore (Model 4). Manufactured by the Densmore Typewriter Company of New York, New York. The Model 4 had 39 keys and 76 characters. Serial number: 12,055. (c.1897) Availability: Scarce. Value: $300.

Densmore (Model 5). Manufactured by the Densmore Typewriter Company of New York, New York. The Model 5 had 43 keys and 84 characters. Serial number: 27,966. (c.1907) Availability: Scarce. Value: $300.

Diplomat Portable

Diplomat Portable. Manufactured by S. Niepperg. Serial number: 200,739. (c.1922) Availability: Common. Value: $125.

Draper

Draper. Manufactured by the Chicago Typewriter Company of Chicago, Illinois. First seen on the market in 1906 and was also sold through the marketing channels of Sears, Roebuck & Company. Serial number: 39,919. (c.1906) Availability: Scarce. Value: $700.

Edelmann

Edelmann. Manufactured by Wernicke Edelmann & Co., Germany. It had a type ball that resembled the Blickensderfer, but it used a lever for type key selection. (c.1897) Availability: Scarce. Value: $1,800.

Edison Margin Justifier

Edison Margin Justifier. Manufactured by Justifier Sales Company of 2022 Glendale Building, Los Angeles, California. An attachment that could be mounted on most typewriters to provide an even right hand margin. Availability: Rare. Value: $500.

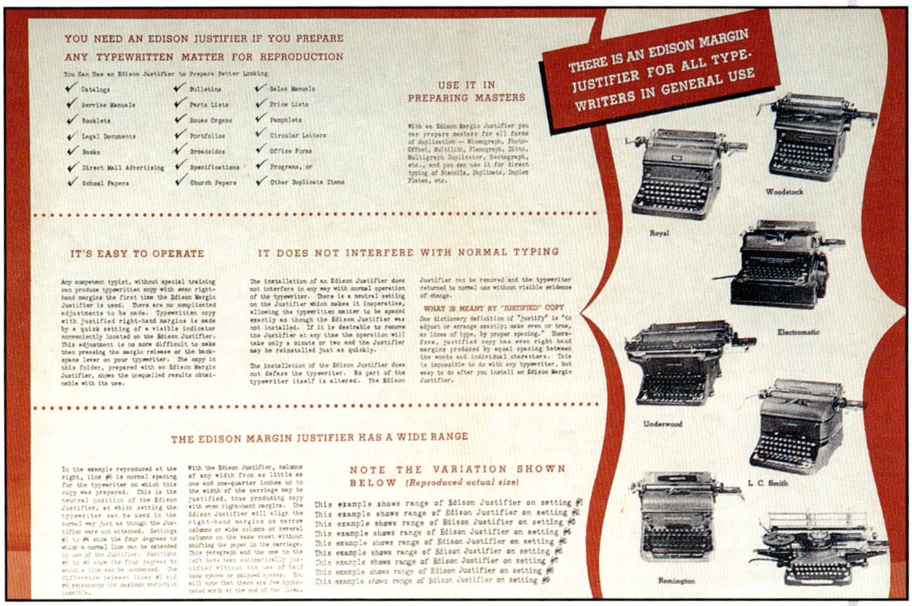

Edison (A.B. Dick)

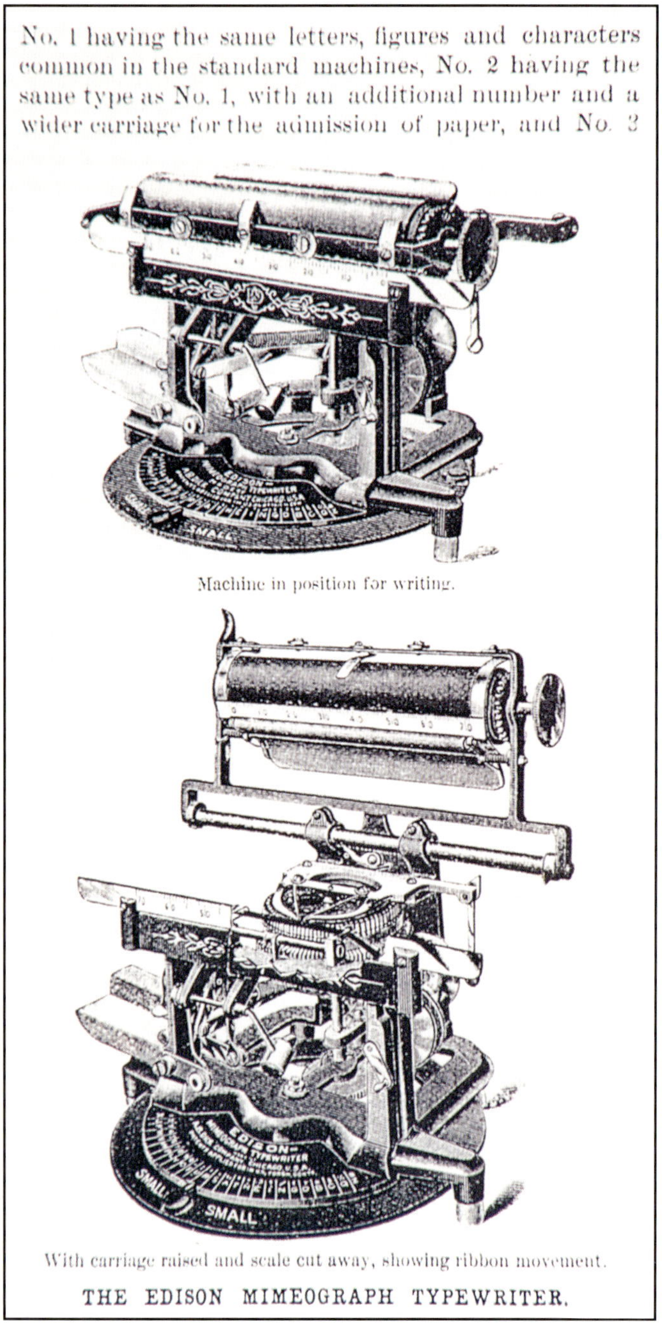
No. 1 having the same letters, figures and characters common in the standard machines, No. 2 having the same type as No. 1, with an additional number and a wider carriage for the admission of paper, and No. 3

Machine in position for writing.

With carriage raised and scale cut away, showing ribbon movement.

THE EDISON MIMEOGRAPH TYPEWRITER.

Edison (Models - 1-2-3). Manufactured by A. B. Dick Company of Chicago, Illinois. The **Model 1** has the same letters, figures and characters as the standard machine (78 characters); the **Model 2** has additional keys and a wider carriage (86 characters); the **Model 3** has characters suitable for writing English, French and German (90 characters). Was designed to cut stencils and can also be used if a high rate of speed is not desired. It was a type bar machine where inking is directly from a ribbon. Advertised as 1 ½ to 2 times faster than handwriting. Patented in 1895 by A. B. Dick. Thomas A. Edison was not involved in its development. It was withdrawn from the market in 1897 because other manufacturers forbade their dealers to sell the Edison. Availability: Rare. Value Each: $5,000.

Edland

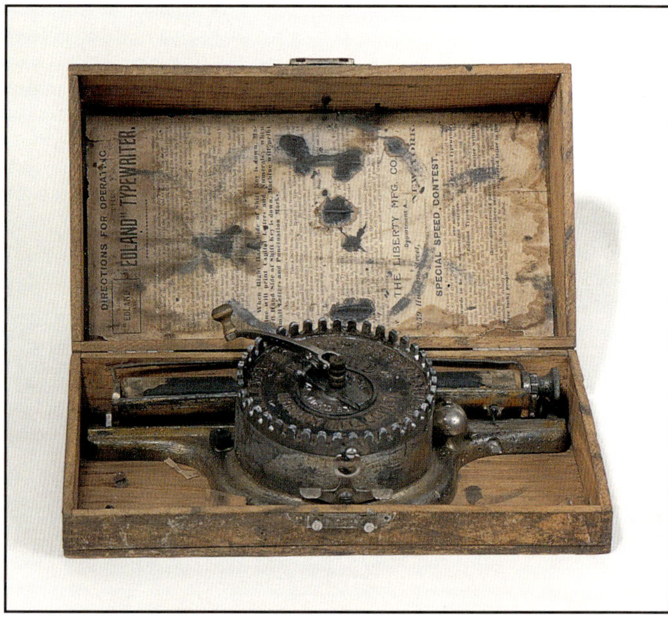

Edland (Model – Engraved). Manufactured by the Liberty Manufacturing Company of New York, New York. Invented by Joseph L. Edland of Brooklyn, New York, and patented on July 14, 1891. There are two models of the Edland. The first model (pictured above) has an engraved circular keyboard. The second model had the keyboard printed on paper and inserted. It was an indicator type, which means it moved in a circular fashion around a three-row wheel. When the desired character was reached, the dial was depressed to print the selected character. (c.1891) Serial number: 190,773. Availability: Rare. Value: $5,000.

Elliott-Fisher

Elliott-Fisher. Manufactured by Elliott-Fisher Company of Harrisburg, Pennsylvania. Invented by Robert L. Fisher. Introduced in 1896 as the Elliott-Hatch typewriter with a down stroke type action designed for a special application; i.e. to type in bound books. This was a useful feature considering the fact that pen and ink were still used for writing in accounting ledgers. In 1903, a mechanism was added to the machine that allowed it to add columns. It was a practical solution to a special application and as a result, many machines were sold. As a matter of fact, the book typewriter was so superior that it survived several mergers and buy-outs. For example, the Fisher Book Typewriter merged with the Elliott-Hatch Book Typewriter and became Elliott-Fisher. Then, in the 1920s, Underwood purchased the Elliott-Fisher Company. Serial number: 11,622. (c.1903) Availability: Scarce. Value: $300.

Elliott-Fisher Underwood. Manufactured by Elliott-Fisher Company. See comments under Elliott-Fisher. The Elliott-Fisher stood on an iron pedestal as show above. Serial number: 190,773. (c.1920) Availability: Scarce. Value: $300.

Ellis

The Ellis was manufactured in both manual and electric versions. It was described in the *Equipment Research Corporation* with the following features: "9 column adding and listing capacity; standard size carriage; combination adding, bookkeeping, and typing with single or double counters; duplex model available with direct subtraction; non-add feature which controlled the carriage so one might imprint a number but not add. (i.e. Bank ABA numbers for wire transfers, etc.); visible printing – available with automatic carriage return and controlling on non-add and non-print features; tab stops of a unique design are set on a bar at the front of the carriage to control the adding and subtracting functions of the counter; the counter on the duplex models provided for accumulating totals in either counter or for transferring the total accumulation from one counter to another; subtraction is direct and may be controlled automatically in definite column position by tabulation of the carriage or manually by means of a key on the keyboard; available as a billing machine combination unit with two counters and subtraction could be adapted."

Ellis (Model - Manual). Manufactured by the Ellis Adding Typewriter Company of Newark, New Jersey. A combination typewriter and adding machine with both a typewriter and adding machine keyboard. Introduced in 1910, several years after the Moon-Hopkins, their major competitor. (c.1910) Availability: Rare. Value: $2,500.

The design strategy of the developers of the Ellis was to provide two separate machines (adding and typewriting) under one frame. Each performed entirely independent of the other. All adding and subtracting functions (numeric functions) were set up on the adding machine keyboard and all typing (alphabetical functions) was done independently of the adding machine on the typewriter. While the actual typing on the typewriter is a manual operation, the carriage return and adding machine stroke mechanisms are electrically driven by an attached motor. The Ellis Combination Duplex Model originally sold for approximately $1,275 and was not considered expensive for that period. The Ellis adding machine and typewriter was the invention of Halcolm Ellis, a resourceful and talented inventor. He subsequently developed the French M.A.P. typewriter.

Emerson

Richard W. Uhlig was one of America's top mechanical artists. His genius was responsible for many typewriter designs and the Emerson ranks as one of his best undertakings. It was certainly one of his most unique. The type bars did not swing in any manner, heretofore, considered to be conventional. Half of the type and type bars were located to the left of the printing point and half were located to the right. As a key top was depressed, the type and type bar would swing at an oblique angle to the point of print. After the factory had been moved to Momence, Illinois, for a few years, it was later sold to Sears, Roebuck & Company. The factory moved again to Woodstock, Illinois. The name was eventually changed to Woodstock and a Woodstock typewriter, the Model 4, replaced the old Emerson. Some models have surfaced associated with Harry A. Smith's name. He was known to have frequently purchased distressed merchandise, and then marketing the product under his own name. It appears that he did the same with the last of the Emerson inventory.

Emerson (Model 3). Manufactured by the Emerson Typewriter Company of Kittery, Maine. Invented by Richard W. Uhlig and patented in 1902. Serial number: 6,166. (c.1907) Availability: Common. Value: $450.

Empire

Empire, The. Manufactured by the Williams Manufacturing Company of Montreal, Canada. This is the Canadian version of the American typewriter called the "Wellington," which was manufactured in Plattsburg, New York. In Germany, it was manufactured under the name "Adler." Serial number: 17,076. (c.1895) Availability: Common. Value each: $125.

Erika

Erika (Model - 3 Bank Folding). Manufactured by Siedel and Naumann of Dresden, Germany. Siedel and Naumann began producing typewriters as early as 1900, with a typewriter they named the Ideal. Later, in 1920, the company produced the Erika 3 Bank, folding model. It appears to have been named after the inventor's daughter. It was also sold under the name Bijou. Siedel and Naumann had a long history of manufacturing business machines, which included calculators. Serial number: 18,494. (c.1910) Availability: Common. Value: $75.

Erika (Model 5 - Tab). Manufactured by Siedel and Naumann of Dresden, Germany. (See comments under Erika Model - 3 Bank Folding.) Serial number: 30,074. (c.1927) Availability: Common. Value: $75.

Erika (Model - S). Manufactured by Siedel and Naumann of Dresden, Germany. See previous comments. Serial number: 569,665. (c.1936) Availability: Common. Value: $75.

Erika (Model 6). Manufactured by Siedel and Naumann of Dresden, Germany. See previous comments. Serial number: 523,685-6. (c.1936) Availability: Common. Value: $75.

Exxon

Exxon Intelligent Typewriter (Model 215 – Series 200). Manufactured by Exxon Office Systems, 101 Gordon Drive, Lionville, Pennsylvania 19353. The serial number of this machine is L2,500,035 which indicates that it may have been the 35th machine manufactured. It is not believed that many were made before Exxon decided to get out of the office equipment business. Availability: Common. Value: $125.

E-Z Writer

E-Z Writer. By an unknown manufacturer, this is the American version of the German Ludolf Circular Index Typewriter. Serial number: None. (c.1930) Availability: Scarce. Value: $150.

Fay Sholes

Fay Sholes (Model 6). Manufactured by the Fay Sholes Typewriter Company. (See Rem Sholes.) Serial number: 22,307. (c.1901) Availability: Scarce Value: $300.

Fay Sholes (Model 7). Manufactured by the Fay Sholes Typewriter Company. (See Rem Sholes.) (c.1905). Availability: Scarce. Value: $300.

Fitch

Originally patented by Eugene Fitch of Des Moines, Iowa, in 1886. Initial models of the Fitch had a two-row keyboard, but later models had three rows. It had a down stroke type action with the type directly behind and above the platen. Printing was visible. A curved basket in front of the machine stored paper. The type itself was made of hard or vulcanized rubber. After a brief and unsuccessful period of manufacture in the United States, an English version was made in London, England. It was redesigned but it also met with little success. The primary reasons for the lack of success of the Fitch was an unconventional keyboard. The arrangement of the keys did not lend themselves to faster typing. The inking system was an awkward and messy one. Although the Fitch typewriter was not successful in its day, it is one of the most desired by collectors today. It commands a premium price because it is not only unique in design, but also rare.

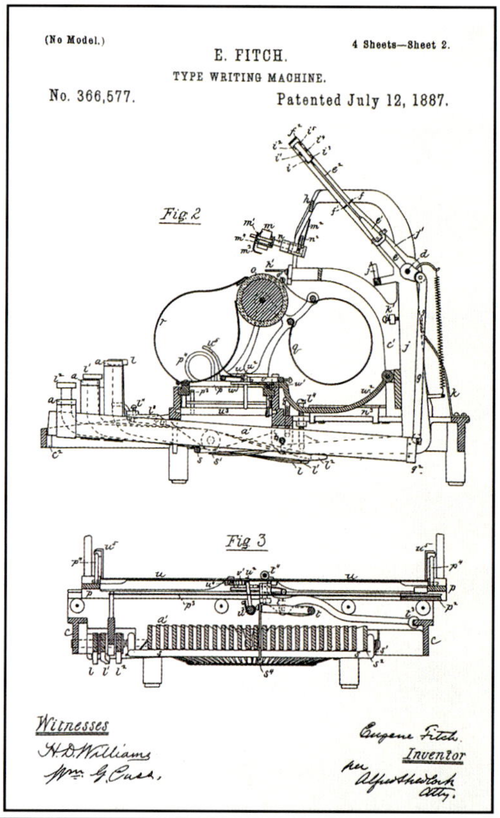

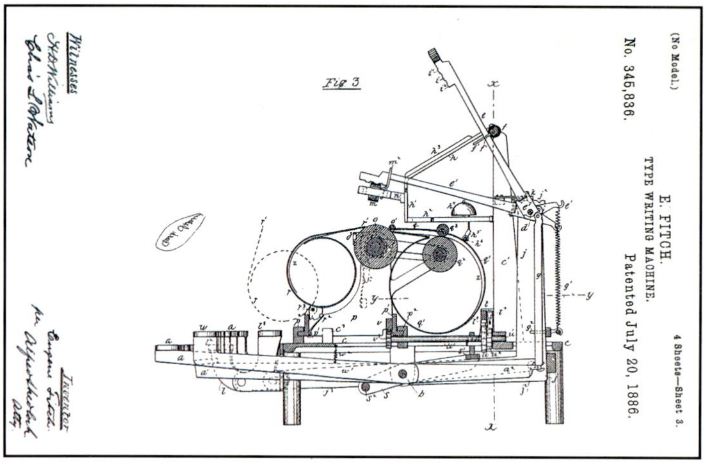

Fitch (American Version). Manufactured by Brady Manufacturing Company of Brooklyn, New York. Has a two-row keyboard. Serial number; 323. Availability: Rare. Value: $15,000.

Fitch (English Version). Manufactured by Brady Manufacturing Company of Brooklyn, New York. Has a three-row keyboard. Serial number: 3,293. Availability: Rare. Value: $16,000.

Florida

Florida Portable. Manufactured in Spain. Platen length: 9-1/8" with 44 keys. Availability: Common. Value: $50.

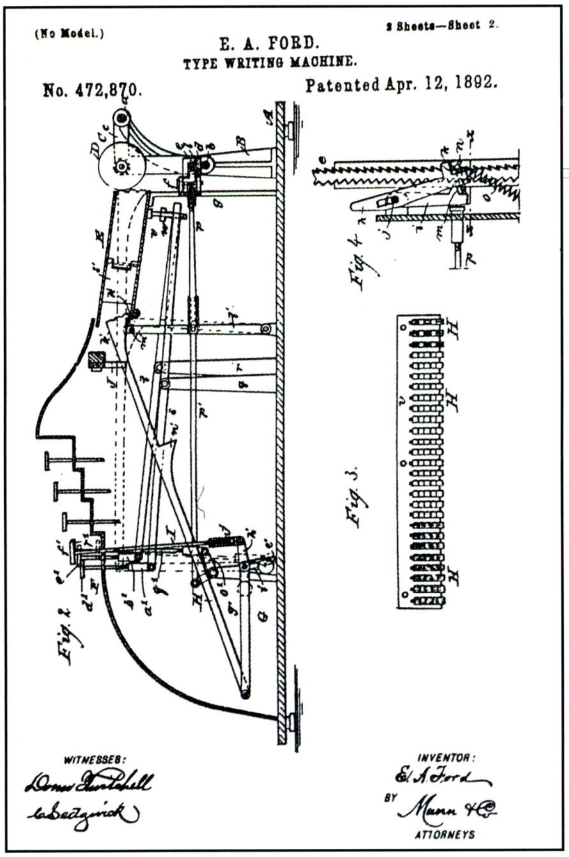

Ford

Ford (Model 1 – Cast Iron). Manufactured by the Ford Typewriter Company of New York, New York. The Ford was a thrust action, lightweight, visible writing, double shift machine. It is another elaborately designed machine that commands respect and price with collectors today. Although considered rare, they still occasionally surface in auctions and other channels. The first model (pictured above) has a cast iron frame. The second model is aluminum. They weighed 16 and 11 pounds respectively. There is said to be no connection between the Ford Typewriter Company and the Ford Motor Company. Serial number: 860. (c.1895) Availability: Rare. Value: $16,000.

Ford Pitt

Fort Pitt. Manufactured by Daugherty Typewriter Company. The Fort Pitt was a variation of the Pittsburg Model 11. It was also sold as the Reliance Visible, Shilling, Broadway and the American. Serial number: 61,053. (c.1912) Availability: Scarce. Value: $400.

Fox Typewriter Factory.

Fox

The Fox Typewriter Company of Grand Rapids, Michigan, manufactured the Models 1, 2, and Sterling Portables. The folding portable resembled the Corona Folding Typewriter (both in appearance and principle) so much, that the Corona Typewriter Company filed suit. The Fox folding portable line was removed from the market and eventually was replaced by the Fox Sterling, which lasted only a year before the company went out of business.

W. R. Fox – President of Fox Typewriter Company.

Fox Portable (Model 1). Serial number: 3,108. (c.1917) Availability: Scarce. Value: $600.

Fox Portable (Model 2). Serial number: 11,371. (c.1918) Availability: Scarce. Value: $300.

Fox Portable (Sterling Model). Serial number: 312,906. (c.1920) Availability: Rare. Value: $1,000.

The Fox Blindwriters (Models 2 - 3 - 4 and 10) were manufactured by the Fox Typewriter Company of Grand Rapids, Michigan. A blindwriter, under stroke machine patented by William R. Fox and Glenn J. Barrett in 1892. The Model 2, 3, 4 and 10 were upstroke machines similar to the Remington Sholes & Glidden that was originally introduced in 1873. There were at least eight Fox patents between the first Fox and the Model 10. Patents: March 29, 1902; November 29, 1898; September 26, 1899; August 28, 1900; September 4, 1900; August 28, 1901; April 23, 1901, and April 15, 1902.

Fox Blindwriter (Model 2). A 44-key keyboard with an 8.75" platen. This model has a pound sign and degree symbol as well as six fraction symbols. Patents engraved on the machine indicate the first Fox patent was March 29, 1892 and not March 29, 1898 as many have reported in the past. Serial number: 2,508. (c.1901) Availability: Rare. Value: $800.

Fox Blindwriter (Model 3). Serial number: 5,320. c.1900. Availability: Scarce. Value: $700.

Fox Blindwriter (Model 4). Serial number: 25,051. (c.1902) Availability: Scarce. Value: $500.

Fox Blindwriter (Model 10). Serial number: 7,681. (c.1902) Availability: Scarce. Value: $500

Fox Visibles (Models 23 – 24 - 25 & 26) were manufactured by the Fox Typewriter Company of Grand Rapids, Michigan. In 1906, Fox introduced their version of a visible front stroke machine beginning with the Model 23. Several models followed with minor variations.

Fox Visible (Model 23). Serial number: 31,970. Availability: Common. Value: $175.

Fox Visible (Model 25). Serial number: 34,260. Availability: Scarce. Value: $250.

Fox Visible (Model 24). Serial number: 109,688. Availability: Common. Value: $175.

Fox Visible (Model 26). Serial number: 43,770. (c.1904) Has a wide carriage. Availability: Scarce. Value: $250.

Franklin

The Franklin Typewriter Company of Boston, Massachusetts, manufactured the Franklin. Wellington Parker Kidder of Jamaica Plain, Massachusetts, was the inventor. Kidder earlier invented another typewriter mechanism that resulted in the Wellington typewriter. The Franklin has a down stroke action where the type bars rested in a standing semicircle in front and above the platen. In 1907, the Victor Typewriter Company took over the Franklin and released their front stroke Victor typewriter. The Franklin Typewriter Company did not number all of their different models sequentially. This makes it necessary for a collector to rely on the serial numbers and old advertising to determine a sequence of the various models. Richard Dickerson, a collector in California, has researched the Franklin typewriter and has published a comprehensive article in the *ETCetera* newsletter, the most knowledgeable information on the subject to date.

In the author's collection, the following machines are arranged in numerical sequence by serial number. This will give the reader a better idea of the approximate time of the production of the various models during the thirteen-year period prior to the take over by Victor.

Franklin, The (Model - Medallion). Serial number: 618. (c.1891) Availability: Rare. Value: $1,500.

Franklin, The. Serial number: 3,248. (c.1893) Availability: Rare. Value: $1,200.

Franklin, The New. Serial number: 6,611. (c.1895) Availability: Rare. Value: $1,000.

Franklin (Model 7). Serial number: 12,860. (c.1897) Availability: Scarce. Value: $700.

Franklin (Model 8). Serial number: 14,563. (c.1899) Availability: Scarce. Value: $800.

Franklin (Model 9). Serial number: 17,136. (c.1900) Availability: Rare. Value: $1,000.

Friden Flex-o-writer

Friden Flex-o-writer (Model FTCP-8 with Postage Meter). Manufactured by the Recorder Commercial Controls Corporation of Rochester, New York. (c.1955) This was an automatic typewriter that used a paper tape for its recording media. Major application was repetitive letters. Availability: Scarce. Value: $300.

Frolio

Manufactured by Gundka Werke, G.m.b.H of Germany. See Gundka. Invented by Paul Muchajer. This machine was marketed in basically the same design under several names and in numerous countries.

Frolio (Model 5). Serial number: 65,760. (c.1924) Availability: Common. Value: $150.

Frolio (Model 7). Manufactured by Frolio Italiana "Machina Da Scrivere Portatile." Sec. Imex Verona. Serial number: 85. Availability: Common. Value: $150.

G & K

G & K. Manufactured by Paul Muchajer, Hersteller: Gundka-Werk, G.m.b.H., Kurstr., 14 Brandenburg, Havel. (c.1924) Availability: Common. Value: $125.

Geniatus

Geniatus. A swinging sector index type machine that used a rubberized belt on which the type was located. It is believe to be of German origin. Serial number: 7,212. (c.1924) Availability: Scarce. Value: $300.

Gisela

Gisela. Manufactured by the Gisela Typewriter Works, Günther & Company, Berlin 027, Schellingstr. 14. Rare German machine manufactured in 1921. It was a double shift machine with 30 keys and a 9.75" platen. Serial number: 114. Availability: Rare. Value: $1,500.

Gourland

Gourland. Manufactured by the Gourland Typewriter Corporation at 120 Broadway in New York, New York. Their factory was in Brooklyn, New York. Most of the credit for bringing this product to market has been given to M. J. Gourland, but Charles Spiro is said to have played a part in its design. Moving the paper guide fingers to the edges of the paper set the margins. Serial number: 1,566. (c.1920) Availability: Scarce. Value: $250.

Grandjean

Grandjean (Model - Steno). Manufactured by S.A. Stenotype Granjean, 8, rue St. Augustin, Paris 2me. Invented by Marc Granjean. This was the model that eventually would become the Stenotype machine. Serial number: 12,614. (c.1910) Availability: Common. Value: $125.

Grandville

Grandville Automatic. Manufactured by the Grandville Manufacturing Company of Providence, Rhode Island. Invented and patented by Bernard Grandville of Dayton, Ohio, who played a part in the invention of the Rapid Typewriter. Both efforts were on the "thrust type action." It was first to be manufactured by the Grandville Manufacturing Company and then by the Mossberg & Grandville Manufacturing Company, both of Providence, Rhode Island. The side to side carriage movement was controlled by a key-top on each side of the typewriter. Serial number: 5,964. (c.1896) Availability: Rare. Value: $5,000.

Gritzner

Gritzner. A typewriter designed to type on flat surfaces, i.e. maps and blue prints. Type: 4-1001-002. Serial number: 4,572. (c.1966) Availability: Scarce. Value: $300.

Groma

Groma (Model - N). Manufactured by Mechanik, Groma,Veb, Markersdorf /Chemnitzal. Serial number: 273,063. (c.1924) Availability: Common. Value: $125.

Gundka

The following machines (Gundka Model 5, Frolio Models 5 and 7 [see these under Frolio previously], G & K, M & W and the Write-Easy) were versions of the original patent of 1924 by Paul Muchajer under the Gundka name and marketed in various countries under different names.

Gundka (Model 5). Serial number: 57,613. Availability: Common. Value: $125.

M & W. Availability: Common. Value: $125.

Write-Easy (Gundka Type). Serial number: 81,670. Availability: Common. Value: $125.

Hall Index

The inventor of the Hall Index Typewriter was Thomas Hall. His original patent was dated March 1, 1881 (patent number 238,387). According to the inventor, his intent in designing the Hall Index Typewriter was to "reduce the number of parts in typewriting machines and to reduce the size and cost as well." Hall accomplished his mission and gave the American typewriter industry its first successful index machine. The action of the Hall is best described in his 1881 patent and is quoted partially below:

> *"The general plan of this machine includes a base or platen, on which the paper rests in order to be printed. This may be a plate, bar, or roller. Over this the paper is fed, either in sheets or from a roll, with a regulated intermittent motion, to form the lines of writing. This base or platen and the feed-rollers mentioned are a part of the base of the machine. The type-form is movable, both for bringing any given letter into alignment with the opening in said plate from the position of one letter to that of the next. The motion required to bring any given letter into alignment with the opening in the plate is directly impacted by movement of the hand grasping a small lever or arm and guiding the same by means of a pointer thereon fitting definitely arranged holes, any one of which, when the pointer is set there, brings its appropriate letter over the opening, through which it is pressed to print the letter. Further, the motion of this directing arm is accomplished or directly followed by a depression of the type form and its supporting frame and subsequent elevation these movements being necessary in the act of printing a letter. These movements are made to operate an escapement whereby the perforated plate which carries the type form is permitted to move one notch to the right, bringing the type to the space of the paper appropriate for the next letter.*
>
> *The step-by-step movement of the type carrying mechanism from left to right to form the necessary spacing is accomplished by means of a spring or equivalent motor acting in connection with an escapement. This escapement is of special construction, and consists of an elastic arm, which engages with a rack bar, and is released by the movement necessary in printing, but advances a limited distance in the direction of the movement of the carriage before again engaging the said rack bar. After its advance and engagement the carriage is moved up to it as to a stop the distance of its advance, and this distance may be varied to vary the spacing."*

There were three models of the Hall typewriter. The first two models were based on Hall's original patent of 1881. The third model, with Boston shown on the nameplate was advertised an as "Improved Model" and was manufactured based on the second patent dated February 16, 1886.

Hall (Model 1 – New York). Serial number: 2,004. (c.1881) Availability: Scarce. Value: $600.

Hall (Model 2 – Salem). Serial number: 7,603. (c.1883) Availability: Scarce. Value: $500.

Hall (Model 3 – Boston). Serial number: 12,897. (c.1881) Availability: Scarce. Value: $500.

Hall

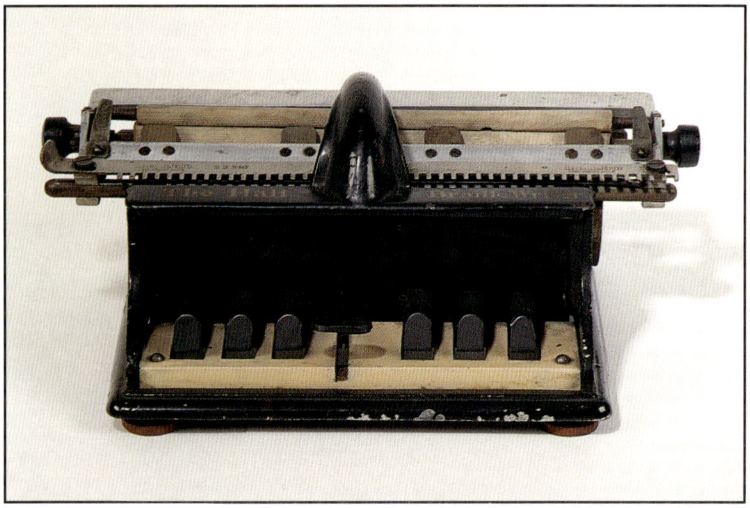

Hall (Model - Braille). Frank N. Hall, not related to Thomas Hall who invented the Hall Index machine, developed this typewriter which had a piano like keyboard. By depressing one of the six black keys on the keyboard, a braille cell was embossed. Serial number: 22,310. (c.1891) Availability: Scarce. Value: $250.

Hamilton - *See Automatic*

Hammond

James B. Hammond was the inventor of this remarkable typewriter. His background included an education that would have prepared him for the ministry. He was also a Civil War reporter for the *New York Tribune* as well as a court reporter. This experience may have led him to conceive the idea of developing a typewriter, which maintained the same even impression no matter how hard you struck the keys. This of course, would result in a better appearing final copy.

Hammond also felt that the arrangement of the keys as developed by Christopher Latham Sholes on the Remington was awkward. So he developed his own keyboard arrangement and placed the keys so an operator's fingers could attain greater speed. He named it the "Universal" keyboard. His very first model introduced his keyboard. Of all the wonderful features on his typewriter, his innovative keyboard was not a success. Already thousands of operators had learned the Sholes Standard keyboard (also known as the QWERTY keyboard) and were reluctant to change. The Sholes keyboard remains popular to this day and is still used on our modern day computers.

Fortunately for Hammond, his typewriter had other features that would insure its future success and ultimately make him a millionaire.

These features included the following:

- Interchangeable type,
- Variable spacing between characters,
- Automatic type impression,
- The ability to easily change type styles to other available ones for symbols or a foreign language.

Hammond died on January 27, 1913. His typewriter survived his death and continued to be manufactured for many years.

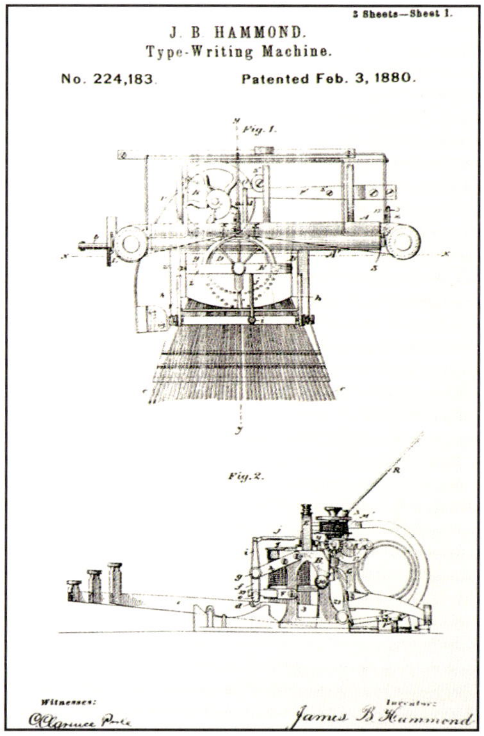

Hammond (Model 1). Manufactured by Hammond Typewriter Company at 69th Street and East River in New York, New York. His first model that introduced his Universal keyboard. Serial number: 6200. (c.1885) Availability: Rare. Value: $1,500.

Hammond (Model 1b – Curved Keyboard). Manufactured by Hammond Typewriter Company at 69th Street and East River in New York, New York. Serial number: 19,512. (c.1888) Last patent: December 18, 1888. Availability: Rare. Value: $800.

Hammond (Remodeled Version – Curved Keyboard). Manufactured by Hammond Typewriter Company at 69th Street and East River in New York, New York. Serial number: 6,332. (c.1893) Last patent: July 18, 1893. Availability: Rare. Value: $800.

Hammond (Model - Curved Keyboard). Manufactured by Hammond Typewriter Company at 69th Street and East River in New York, New York. This model had a wood lever for a backspace that was located directly under the keyboard and to the left. Serial number: 38,353. (c.1905) Availability: Scarce. Value: $400.

Hammond (Model 12 - Straight Keyboard). Manufactured by Hammond Typewriter Company at 69th Street and East River in New York, New York. Serial number: 83,036. (c.1905) Availability: Common. Value: $250.

Hammond (Multiplexor Model – Straight Keyboard). Manufactured by Hammond Typewriter Company at 69th Street and East River in New York, New York. Serial number: 219,254. Availability: Common. Value: $125.

Hammond (Multiplexor Model – Curved Keyboard). Manufactured by Hammond Typewriter Company at 69th Street and East River in New York, New York. Serial number: AL111, 393. Availability: Scarce. Value: $450.

Hammond (Model - Folding). Manufactured by Hammond Typewriter Company at 69th Street and East River in New York, New York. Serial number: F235,325. (c.1926) Availability: Common. Value: $125.

Harris

Harris Visible (Model 4). Manufactured by the Harris Typewriter Company of Fond du Lac, Wisconsin. A no frills, front stroke typewriter that was designed by D. C. Harris and marketed by Sears, Roebuck & Company of Chicago, Illinois. In 1914, the Rex Typewriter Company produced this machine under the names: Aristocrat, Reporter's Special, Rex, and Rex Visible. Later the Rex Typewriter Company also manufactured the Demountable Typewriter. Serial number: 11,784. (c.1911) Availability: Common. Value: $75.

Harris Visible (Model 5). Manufactured by the Harris Typewriter Company of Fond du Lac, Wisconsin. See Harris Visible Model 4. Serial number: 21,128. (c.1913) Availability: Common. Value: $100.

Hartford

Hartford (Model 2). Invented by John M. Fairfield and had a double keyboard and an upstroke action. As in the case of the Caligraph, there was a key for each character. The model pictured above appeared on the market around 1897. There was also a Model 3 that appeared in 1905. When the factory moved from Hartford, Connecticut to Cleveland, Ohio, the name was changed accordingly from the Hartford to the Cleveland. Serial number: 5,693. (c.1894) Availability: Rare. Value: $2,500.

Helios-Klimax

Helios-Klimax. Manufactured by Helios-Schreibmaschinen-Fabrik, D.R.P. Berlin S.42 Ritterstr. 123. Specialabteilung der firma/nev. Metallwarenfabrik gegründet 1869. The Helios has 21 keys with a double shift. It had a single element type, which is interchangeable, allowing for a variety of type styles. Serial number: 14,612. (c.1914) Availability: Scarce. Value: $2,500.

Hermes

Hermes (Model - Ambassador). Manufactured by Paillard S. A. Yverdon, Switzerland. Platen length: 13" with 46 keys. Carbon and fabric ribbon. (c.1960) Availability: Common. Value: $75.

Hooven Automatic Typewriter

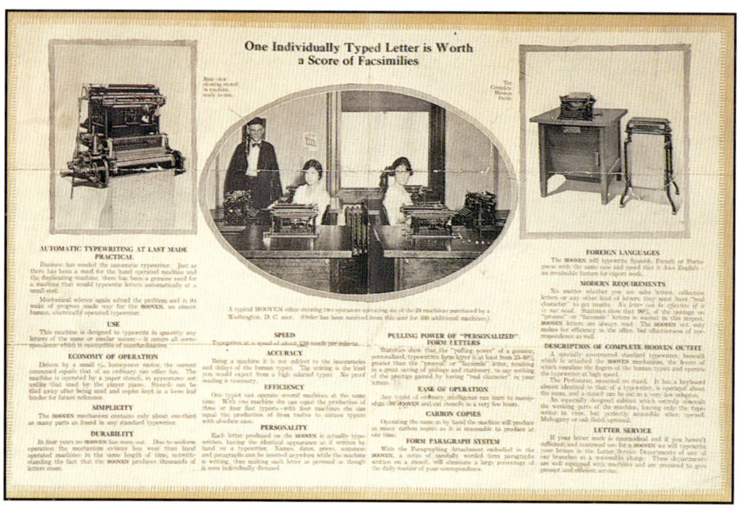

Hooven Automatic Typewriter. Manufactured by the Hooven Automatic Typewriter Company, Hamilton, Ohio. The purpose of the Hooven Automatic was to produce individual letters that looked like they were personally typed by an operator, but in fact, were typed mechanically with the help of a mechanism resembling a player piano. A perforator (pictured above) punched the roll of paper. A typewriter, usually an Underwood with linkage attached to the type bar, would play each letter out at very rapid speeds. It was similar to the Autotypist and the Robityper. (c.1920) Availability: Rare. Value (complete unit): $2,000.

I B M

I B M (Model 121 – Electric). Manufactured by the IBM Corporation. Serial number: 23,938. Availability: Common. Value: $150.

Ideal

Ideal (Model - A). Manufactured by Siedel & Naumann of Dresden, Germany. Although made in Germany, an American, E. E. Barney, patented the Ideal in 1897. Barney was affiliated with the Remington Typewriter Company and was in charge of their Research & Development. The first Ideal was introduced in 1900 with the second model (pictured above) appearing in 1902. Serial number: 1,001,268. (c.1902) Availability: Scarce. Value: $400.

Ideal (Model - B of 1913). Manufactured by A.G. Vorm. Seidel & Nauman. Platen length: 12.5" with 45 keys and a 5-key tabulator. Availability: Common. Value: $175.

Ideal Stencil Machine

Ideal Stencil Cutter (Model 3). Manufactured by the Ideal Stencil Machine Company of Belleville, Illinois. A circular type stencil cutter that had as its main feature an automatic carriage that guaranteed even spacing between character, words and line spacing. There were three models and each model provided letters and figures in different sizes: the Model 1 - 1", the Model 2 - .628" and the Model 3 - .4375". Serial number: 8,250. (c.1928) Availability: Common. Value: $50.

Imperial

Imperial (Models – B and D). Manufactured by the Imperial Typewriter Company of Leicester, England. The Imperial Typewriter Company's beginnings go back to 1902. However, the Model D appeared shortly after World War I. Platen length: 9.75" with 43 keys. Serial number: 663,836. (c.1919) Availability: Scarce. Value each: $600.

Imperial (Model 1 – Good Companion). Manufactured by the Imperial Typewriter Company, Ltd. of Leicester, England. Serial number: P812 EC 033. (c.1932) Availability: Scarce. Value: $100.

International

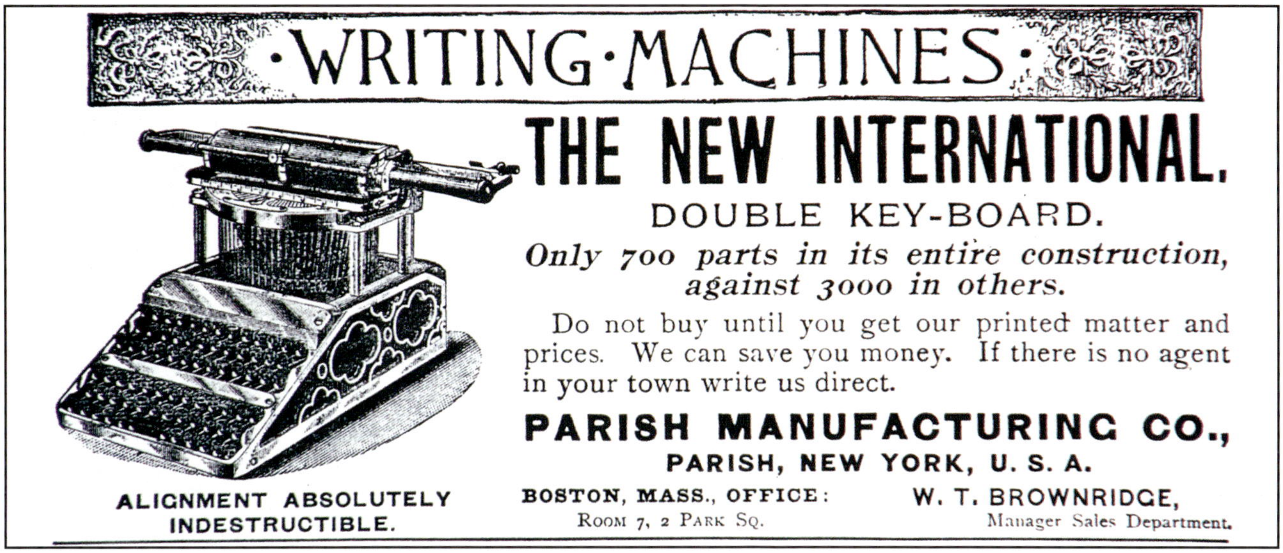

International (Model 2). Manufactured by the International Typewriter Company of Parish, New York. This double universal keyboard typewriter was invented by Lucien Stephen Crandall and patented in 1889. The earliest model was a three-bank under stroke machine. The machine pictured above is the second model. Notice the unique and beautifully carved cut outs on the side of the base. (c.1889) Availability: Rare. Value: $18,000.

Jewett

Manufactured by the Jewett Typewriter Company of Des Moines, Iowa. George A. Jewett developed and brought to market the Jewett typewriter. Originally called the Duplex and later the Duplex-Jewett and finally, simply Jewett. The early machines were blindwriters. The operator needed to raise the carriage to view what had been typed. It also had a double keyboard.

Jewett (Model 4). Manufactured by the Duplex-Jewett Typewriter Company. Serial number: 11,784. (c.1897) Availability: Scarce. Value: $500.

Jewett (Model 2). Serial number: 5,778. (c.1895) Availability: Rare. Value: $1,000.

Jewett (Model 3). Manufactured by the Duplex-Jewett Typewriter Company of Des Moines, Iowa. Serial number: 8512. (c.1896) Availability: Scarce. Value: $600.

Junior

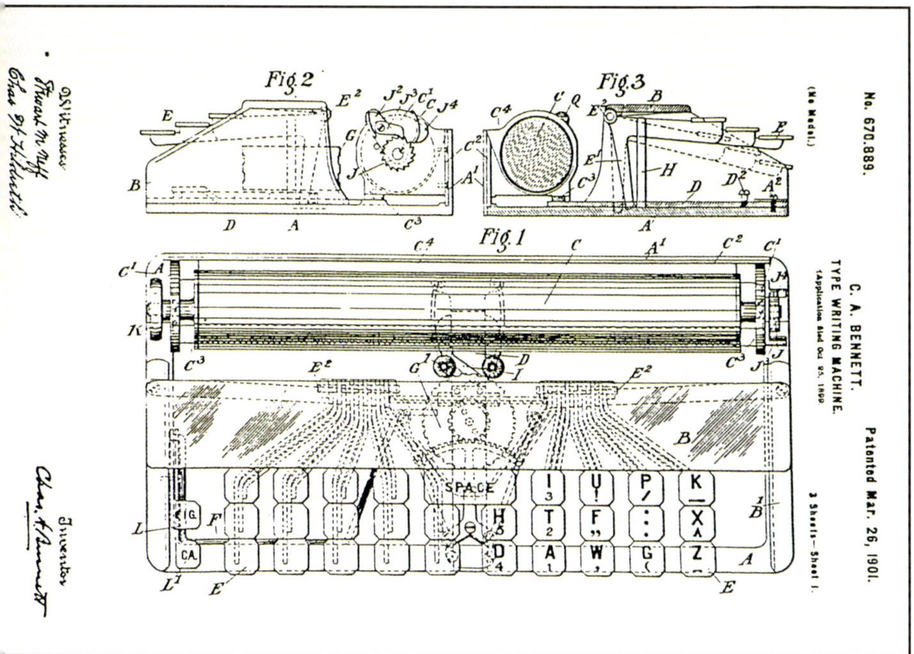

Junior (Model - Bennett Type). This portable typewriter is one of the truly unique antique typewriters still available to collectors. It was first patented on March 26, 1902 by Charles A. Bennett and assigned patent number 670,889. It was rectangular in shape and had a rigid leather cover that could be removed by pulling on two rings located at both ends of the case. The Junior had an unusual three-row keyboard. By depressing one key, the entire keyboard moved down with the key. It was produced in two versions. A common black model and a rare aluminum model. Two rollers were used as the inking system and a type element similar to the Blickensderfer. A later patent introduced a ribbon inking system and was also patented by Bennett in 1907 (See Bennett typewriter). Availability Black model: Scarce. Availability Aluminum model: Rare. Value Black model: $300. Value Aluminum model: $450 .

Junior - German

Junior (Model - GSN). German toy typewriter. (c.1920) Availability: Common. Value: $125.

Kanzler

Kanzler (Model 4). Manufactured by Artiengesellschaft fur Schreibmaschinen Indistrie of Berlin, Germany. The Kanzler was a thrust action machine similar to the Adler or the Wellington. It had four rows of keys, a single shift and a curved keyboard. Probably the most unique feature on the Kanzler was the design of the type bars and the keyboard. There were eleven type bars and eight characters per type bar and forty-four keys on the keyboard. Based on which key was depressed one of those eight characters on that particular type bar was aligned to print that letter. There were four models manufactured before the Kanzler was removed from the market. Serial number: 9,587. (c.1903) Availability: Scarce. Value: $700.

Kappel

Kappel. Manufactured by Maschinenfabrik Kappel A-G of Chemenitz – Kappel, Germany. The Kappel Company began operations as early as 1860 but their first typewriter was introduced in 1914. This portable was manufactured about 1936. Serial number: 105,289. Availability: Common. Value: $100.

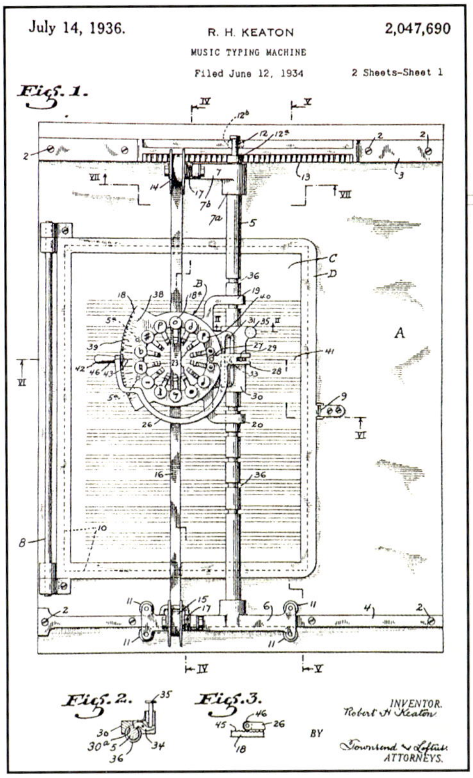

Keaton

Keaton (Model - Musicwriter). Manufactured by the Keaton Music Typewriter Company of San Francisco, California. Robert H. Keaton was the inventor of this music typewriter. Keaton's original patent was granted on July 14, 1936. It only had 14 keys. Patent Number: 2,047,690. His second patent was granted on March 17, 1953 and had 33 keys. Serial number: 1,961. (c.1936) Availability: Rare. Value: $4,000.

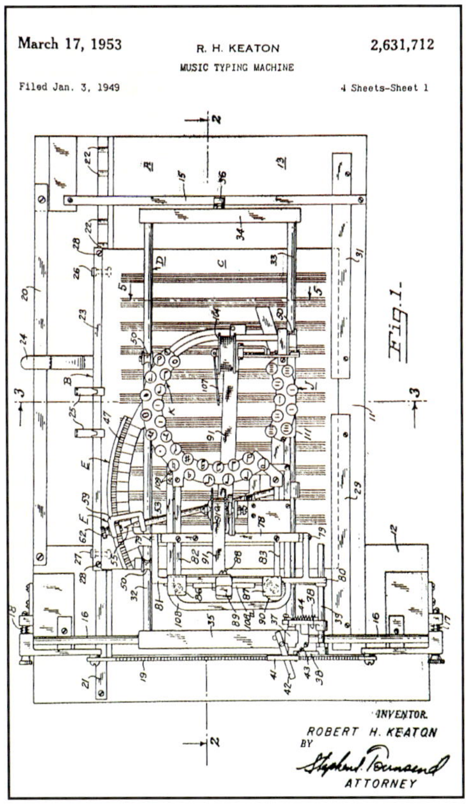

Klein-Adler - *See Adler*

Keystone

Keystone (Model 1). Manufactured by the Keystone Typewriter Company of Harrisburg, Pennsylvania. The Keystone used a "truss wheel", similar to the Crandall. The type was interchangeable and it used a double shift. It had three rows of keys and a reduced number of parts. In an effort to keep the manufacturing costs down, inexpensive parts and materials were used. As an example, the carriage rail was made of inferior white metal and then attached to a veneered wood base. Gases from the veneer caused the white metal to deteriorate as time passed. As a result, it is not uncommon to find the Keystone with a warped or damaged carriage rail. Serial number: 1,007. (c.1899) Availability: Rare. Value: $3,000.

Lambert

The inventor of the Lambert Typewriter was Frank Lambert. His typewriter was a departure from what had been offered on the typewriter market up to 1896. The keyboard was circular and resembled an old rotary dial telephone. To operate the 28-key keyboard with 84 characters, you depressed the key you wanted to type, which then caused the entire keyboard to tilt and the hard rubber type made contact with the paper. The two models of the Lambert that have surfaced to date are pictured (the Model 1 – Engraved and the Model 2 – Stenciled). The Lambert was also released as the Butler. The four patents attributed to Frank Lambert were 306,761; 607,270; 640,208 and 693,278.

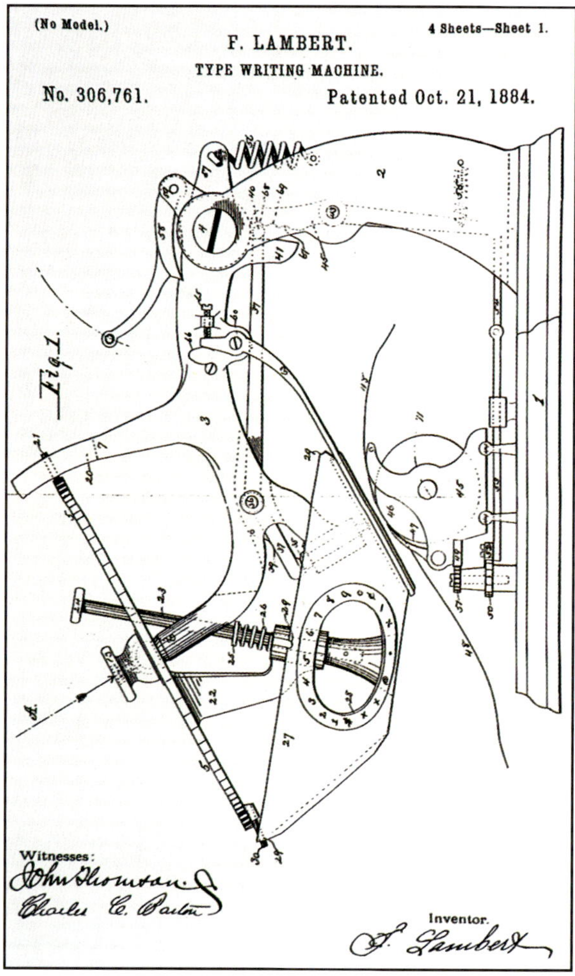

(No Model.) 4 Sheets—Sheet 1.

F. LAMBERT.

TYPE WRITING MACHINE.

No. 306,761. Patented Oct. 21, 1884.

Lambert (Model 1 – Engraved). Manufactured by the Lambert Typewriter Company at 79-83 Washington Street in Brooklyn, New York. Patent number: 306,761. Serial number: 202. (c.1895) Availability: Scarce. Value: $800.

Lambert (Model 2 – Stenciled). Manufactured by the Lambert Typewriter Company at 79-83 Washington Street in Brooklyn, New York. Serial number: 4,148. (c.1902) Availability: Scarce. Value: $700.

Lambert (Model - Butler). Another version of the Lambert Model 2. Serial number: 6,976. (c.1902) Availability: Scarce. Value: $700.

Leta-Con

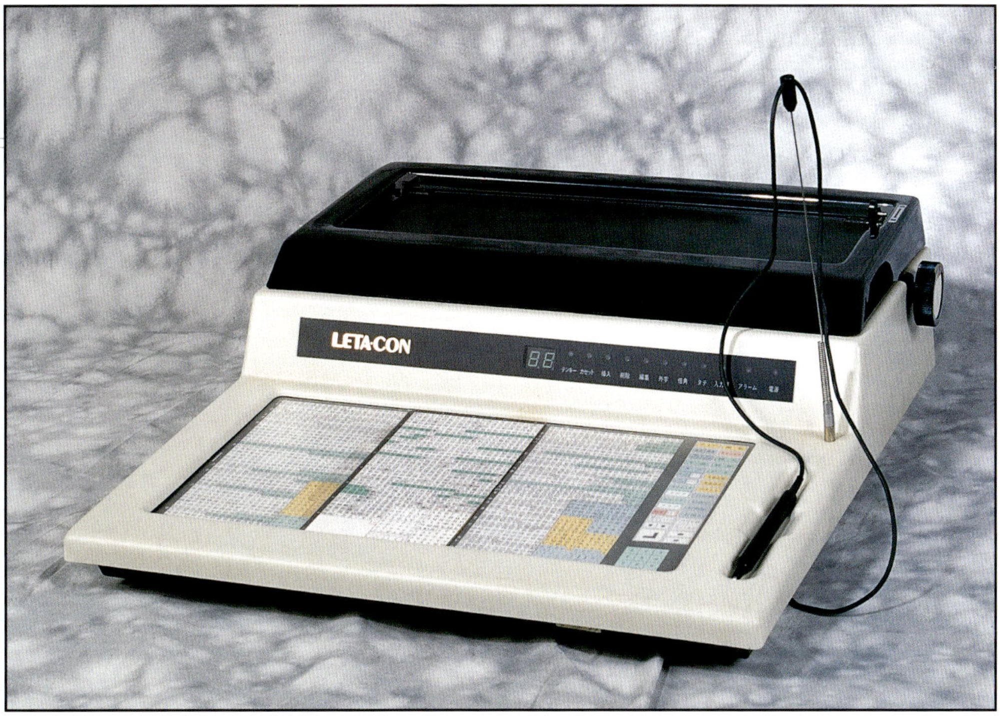

Leta-Con (Model Wpt-1). Manufactured by the Pentel Company of Japan. A Japanese electronic typewriter. Serial number: 81,906. Availability: Scarce. Value: $800.

L. C. Smith & Brothers

The four founders of the L. C. Smith & Brothers Company.

L. C. and W. L. Smith were the first of four brothers to become interested in manufacturing typewriters. Starting in the fall of 1885, they cooperated with Mr. A. T. Brown in the invention and development of a new typewriter with a double keyboard. In 1888, they formed a partnership. The Smith Premier Typewriter Company was formed in 1893. This company later became part of the Union Typewriter Company.

L. C. Smith and Brothers recognized the potential for a typewriter with "visible writing" as opposed to the traditional "blind writers." As a result, the Smith Brothers incorporated the L. C. Smith Company on January 27, 1903. On November 24, 1904, the first "writing in sight" L. C. Smith, their Model 2, was shipped from the factory to Toronto, Canada. Approximately 1,000 machine were shipped within a short time. Other breakthrough features of the Smith typewriter included a segment shift, interchangeable platens and a bi-chrome ribbon attachment with stencil cut out and in-built tabulator.

The first Model 1, was released after the Model 2 on February 2, 1905. The only difference between the two machines was that the Model 1 had 76 characters as opposed to the Model 2, which had 84 characters. Both models were first made with a pivot bar mechanism. The founders recognized this was a weakness and searched for other options. They discovered that "artificial leg" manufacturers were using a new mechanical principle called "ball bearings." The patent for ball bearings belonged to a man named Gardner. The L. C. Smith Company purchased the patent. In 1907, the principle of ball bearings was incorporated into their typewriters. Each type bar used 15 ball bearings. The carriage and type bar segment also incorporated ball bearings.

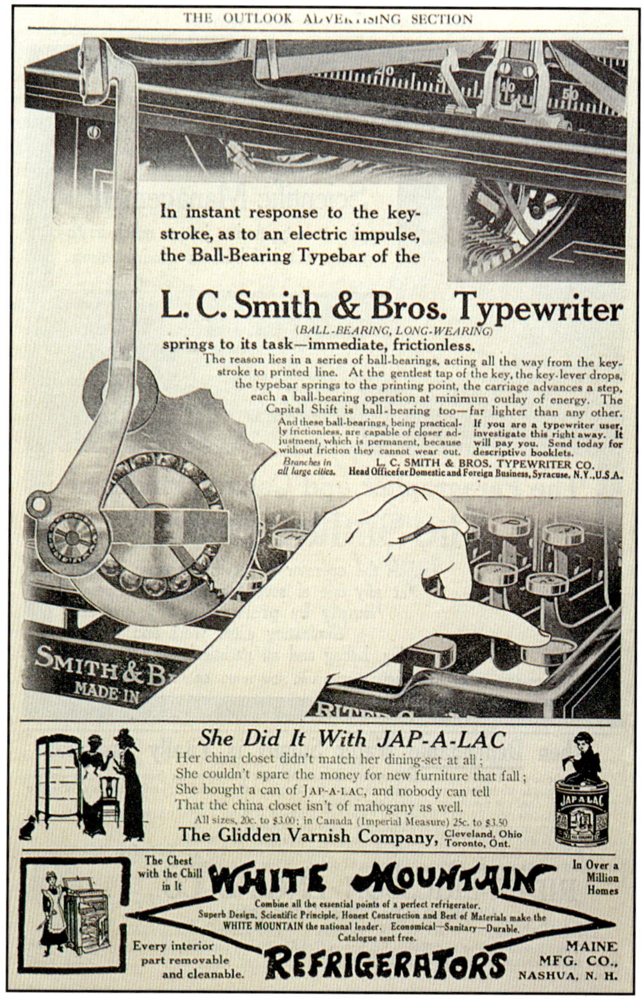

Advertisement for ball bearing type action.

The first model with ball bearing construction was the Model 3 with a 14" carriage. In 1909, the Model 6 with a 20" carriage was made. In 1922, the type bar guide was added and the Model 1 now came to be called the Model 4. The Model 2 became the Model 5.

Mr. Carl Gabrielson. The person responsible for the design and development of L. C. Smith typewriters for over 25 years.

L. C. Smith & Brothers (Model 2). Manufactured by L. C. Smith & Brothers Typewriter Company of Syracuse, New York. The Model 2 was very similar to the Model 1 except that the Model 2 had 42 keys as opposed to 38 keys on the Model 1. Serial number: 12,151-2. (c.1904) Availability: Scarce. Value: $250.

L. C. Smith & Brothers (Model 1). Manufactured by L. C. Smith & Brothers Typewriter Company of Syracuse, New York. Platen length: 10.125" with 38 keys. Serial number: 8,472-1. (c.1905) Availability: Scarce. Value: $300.

L. C. Smith & Brothers (Model 5). Manufactured by L. C. Smith & Brothers Typewriter Company of Syracuse, New York. (See Models 1 & 2 for details.) Serial number: 22,913-5. (c.1911) Availability: Common. Value: $125.

L. C. Smith & Brothers (Model 8). Manufactured by L. C. Smith & Brothers Typewriter Company of Syracuse, New York. This model had a 5-key tabulator system at the top of the keyboard. Platen length: 10" with 42 keys. Serial number: 360,680. (c.1915) Availability: Common. Value: $75.

L. C. Smith & Corona (Model - M-209B-Converter). This was a World War II, U.S. Army Code machine. Dated: April 27, 1942. Serial number: 34,314. Availability: Rare. Value: $4,000.

L. C. Smith & Brothers Standard (Model - Super Speed). Platen length: 10-1/8" with 42 keys. Serial number 1A-1,715,947. (c.1937) Availability: Common. Value: $25

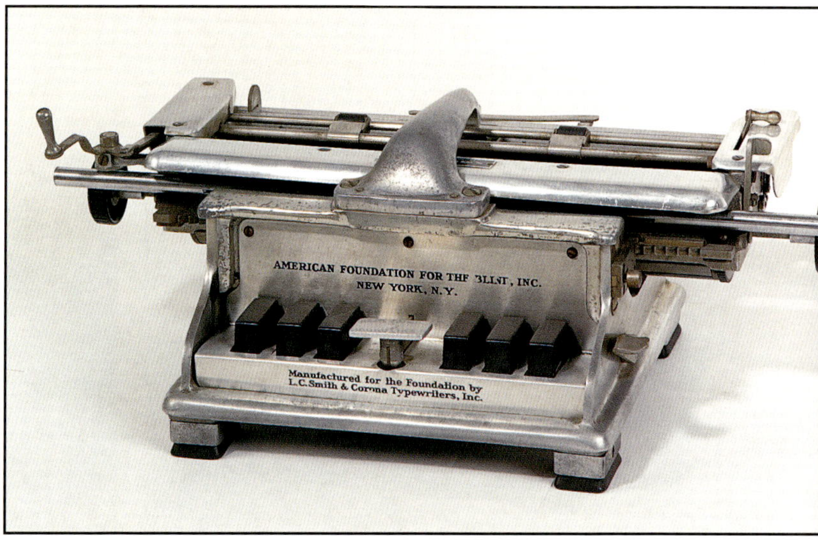

L. C. Smith & Corona (Model - Braille). Manufactured by L. C. Smith & Corona of Syracuse, New York. This braille model resembles the Hall Braille Typewriter of 1891. One name plate on the front carriage reads as follows: "Designed and developed by the American Foundation for the Blind, Inc." Another plate reads: "Licensed by the Underwood-Elliott Fisher Company," and still a third plate reads: "Mfg'd for the American Foundation for the Blind by the L. C. Smith & Corona Typewriter Company." The case is completely nickel-plated and the keyboard as mentioned earlier is identical to the Hall Typewriter with its piano like keys. Serial number: 6-2,54. (c.1926) Availability: Common. Value: $300.

Manhattan

Manhattan. Manufactured by the Manhattan Typewriter Company of Newark, New Jersey. Since the Remington Model 2 was a very successful typewriter, one would think that the Manhattan would have been equally successful, since the they were almost identical. This, however, was not the case. The Manhattan Typewriter Company only lasted a few years. Serial number: 3,641. (c.1898) Availability: Scarce. Value: $175.

Matrix

Matrix (Model - Braille G). Manufactured by the Royal National Institute for the Blind, London, W. I. England. Serial number: 1,870. Availability: Scarce. Value: $300.

McCool

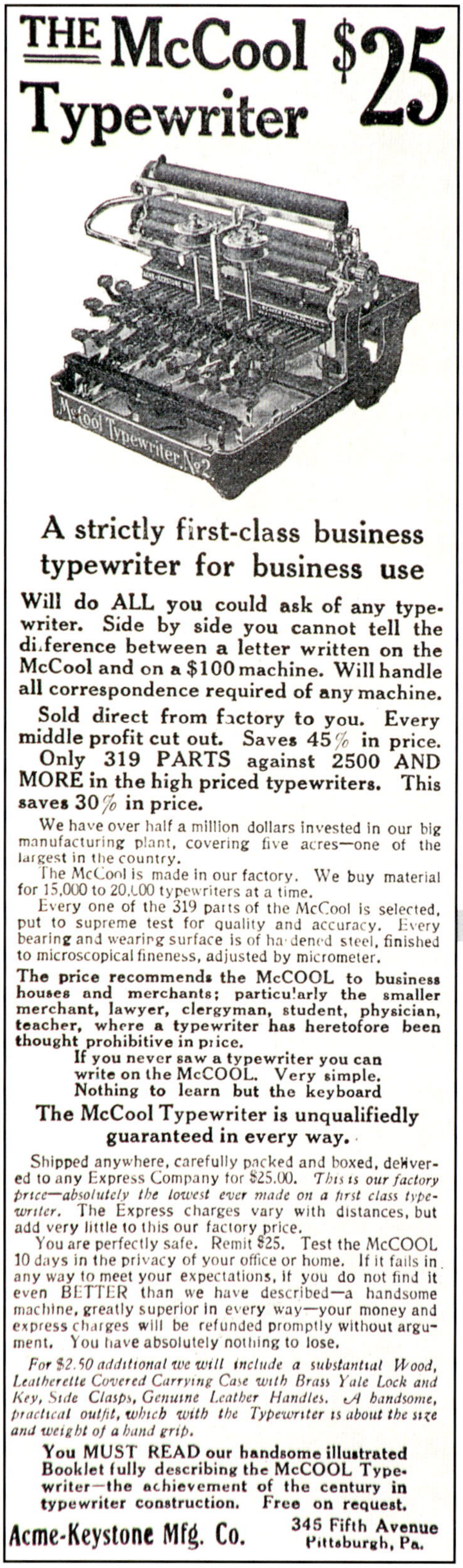

McCool (Model 2). Manufactured by the Acme-Keystone Typewriter Company of Beaver Falls, Pennsylvania. An invention of William A. McCool with only 319 parts. It had a type wheel similar to the Blickensderfer. The McCool was a three-bank machine with a double shift. There were only a few McCool typewriters manufactured. Serial number: 1,650. (c.1910) Availability: Rare. Value: $5,000.

McGreevy

McGreevy Revolving Stamp (Type A). It is a stretch to call this device a typewriter so it was called a "9 in 1 printer." Patent date: April 19, 1920. Availability: Scarce. Value: $50.

Meiselbach

Meiselbach (Sholes). Manufactured by the A. D. Meiselbach Typewriter Company of Kenosha, Wisconsin. Serial number: 4,135. (c.1901) Availability: Rare. Value: $4,000.

Mercantile

Mercantile. Manufactured by the American Typewriter Company of New York, New York. See American Typewriter (type bar model). This is a version of the American typewriter that was manufactured for Montgomery Ward and sold as the Mercantile. Serial number: 14,224. (c.1899) Availability: Scarce. Value: $500.

Mercedes

Mercedes (Model 3). Manufactured by Mercedes Bureau Machinen G.m.b.H. of Charlottenstr., Berlin, Germany. The Mercedes was invented by Franz Schuller. Schuller was a former engineer for Heinrich Kleyer. An electric model was released in 1921. Serial number: 57,047. (c.1920) Availability: Common. Value: $125.

Mercedes (Underwood-Elliott Fisher)

Mercedes (Model - Prima). Manufactured by Underwood-Elliott Fisher Company. Serial number: 94,988. (c.1935) Availability: Common. Value: $100.

Mercedes (Model - Superba). Manufactured by Underwood-Elliott Fisher Company. (c.1940) Serial number: 155,152/1. Availability: Common. Value: $125.

Merritt

Merritt. Manufactured by the Lyons Manufacturing Company of Springfield, Massachusetts. Mortimer G. Merritt of Springfield was granted patent number 510,984 on December 19, 1893. It is described as follows:
"This invention pertains to improvements in the class of type-writing machine in which the platen-roll is mounted on a carriage which has a movement, in the direction of the platen axis, past the common impact point for the characters on the key operating type-bars, and in conjunction with which carriage is an escapement mechanism having connection with each of the key-levers and also the space-bar for imparting the intermittent letter space movement to the platen and its carriage. In this class of machines a devise is usually employed for such a throw-out action, that, at will, the carriage may be run endwise any direction within the limits of its movement." Serial number: 6,899. (c.1893) Availability: Scarce. Value: $1,000.

Mertz

Mertz Universal. Manufactured by Mertz – Werke – Frankfurt A. M. R. Serial number: 57,736. Availability: Scarce. Value: $250.

Meteor

Meteor (Portable). Manufactured by Sachsische Strickmaschinen-Fabrik **Meteor** G.m.b.H., Dresden, Germany. The Meteor was a three-bank keyboard in which the platen and carriage folded in a similar fashion to the American Standard Folding Typewriter. It is reported that the Meteor was produced under the name Pagina around 1922 by the Pagina Fackhaus fur Burobedarf of Berlin . Platen length: 9" with thirty (30) keys and a double shift. (c.1911) Availability: Common. Value: $250.

Mignon

Mignon (Model 2). The Mignon was introduced by the Allgemeine Clektrizitats Gesellshaf Company in Germany about 1904. It is not surprising that they were better known by their initials AEG. Between 1904 and 1928, AEG produced four different models before releasing their type bar machine. It was also produced under other names: Yu Ess, TipTop, Eclipse, and AEG. It eventually became known as Olympia. (c.1904). Availability: Scarce. Value: $600.

Mignon (Model 4). Serial number: 208,202. Availability: Common. Value: $250.

Mignon Special (AEG). Made in Berlin, Germany. Serial number: 249,456. (c.1923) Availability: Scarce. Value: $300.

Molle

Molle (Model 3). Manufactured by the Molle Typewriter Company of Oshkosh, Wisconsin. An invention of John Molle. Also produced under the name Liberty. It was a standard three bank with unusual type linkage. Platen length: 8.75" with 30 keys and double shift. Serial number: 1,160. (c.1913) Availability: Common. Value: $250.

Monarch

Monarch Visible (Model 1). Manufactured by the Monarch Typewriter company of Syracuse, New York. A visible typewriter introduced by the Union Typewriter Company, a trust that was organized to protect pricing and manufacturer profit margins. The author is aware of three models known to exist. They are the Models 1, 2 & 3. The Model 1 had a 9.375" platen and a 38-key keyboard. Serial number: 31,535. (c.1904) Availability: Scarce. Value: $250.

Monarch Visible (Model 3). Refer to Monarch Visible Model 1 for more information. The Model 3 had 42 keys and was available in larger carriage widths, up to and including 26 inches. Serial number: 110,319. (c.1906) Availability: Common. Value: $150.

Monarch Visible (Model 2). Refer to Monarch Visible Model 1 for more information. The Model 2 had a 9" platen and 42 keys. Serial number: 73,049. (c.1906) Availability: Common. Value: $150.

Monarch Portable (Model 5 - Remington). Manufactured by the Remington Typewriter Company of Ilion, New York. This was the same machine as Remington's Model 5 and Streamliner 5 of 1936. Serial number: V837,756. Availability: Common. Value: $50.

Monarch Portable (Pioneer Model - Remington). Manufactured by the Remington Typewriter Company of Ilion, New York. A three and one half row keyboard. The idea was to put eight numbers on four keys so that the operator had to use the shift key to get the 1, 3, 5, and 7. Serial number: C117,141, Availability: Scarce. Value: $175.

Monarch Portable (Model 8 Noiseless - Remington). Manufactured by the Remington Typewriter Company of Ilion, New York. Serial number: E37,228. Availability: Common. Value: $50.

Monarch (Portable). Manufactured by the Monarch Typewriter Company of Syracuse, New York. Serial number: C-66,960. Availability: Common. Value: $50.

Monofix

Monofix. Manufactured by Bauchwitz Pscherer & Company. Serial number: 5,021A. (c.1909) Availability: Scarce. Value: $800.

Montana

Montana Portable. Manufactured by Corso Regio Parco 33, Torino, Italy. Platen length: 9.25", 42 keys. Serial number: 46,600. Availability: Common. Value: $50.

Moon-Hopkins

Top right & right: **Moon-Hopkins (Electric).** Manufactured by the Burroughs Adding Machine Company of Detroit, Michigan. Hubert Hopkins is credited with developing the idea of combining a typewriter and an adding machine for the purpose of generating invoices. A patent on the Moon-Hopkins billing machine was applied for nine years before it was eventually received on September 24, 1912. Although John Dalton financed its early development, Hopkins sold his share to the American Arithmometer Company. After a confusing dispute as to who actually owned the patent, Burroughs (the American Arithmometer Company) finally manufactured the machine during the 1920s. Serial number: 148,986. (c.1920) Availability: Rare. Value: $2,500.

Morris

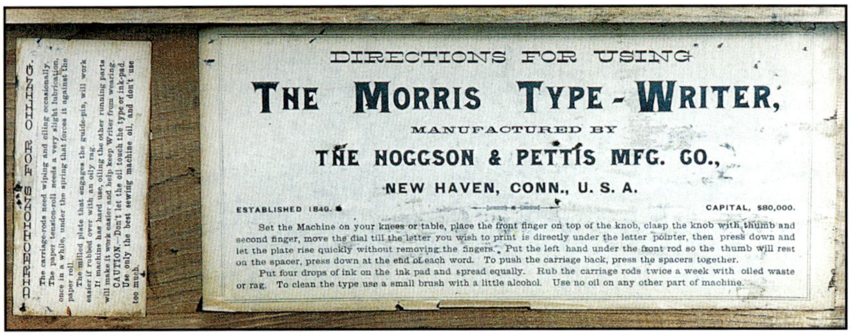

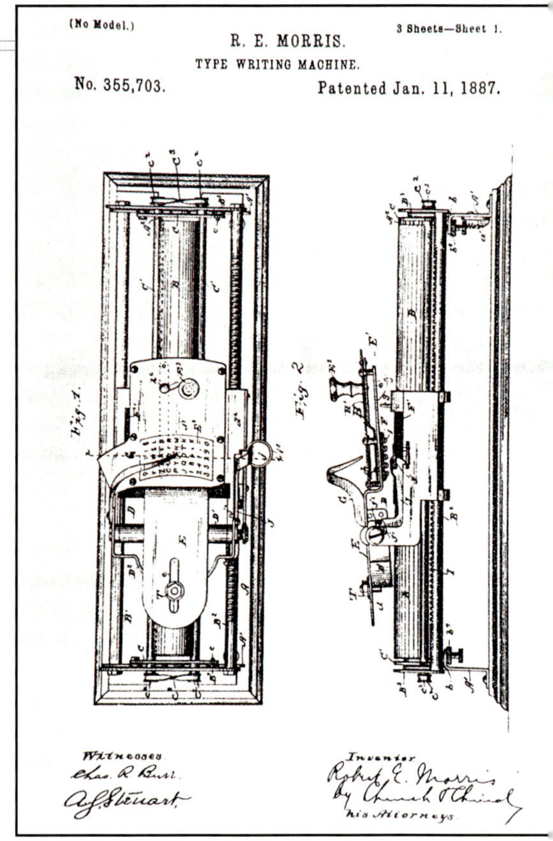

Morris (Model – Upper Case). Manufactured by the Hoggson & Pettis Manufacturing Company of New Haven, Connecticut. Robert Erastus Morris of Wichita, Kansas, was the inventor and patentee. The typewriter was of the index variety and it was advertised in 1899 as being capable of producing 70 words per minute. It is difficult to imagine how one could average or even attain that kind of speed on a device that required the operator to move the top plate so that the desired type face was directly under the pointer. Then one had to grasp the knob and depress it to enable the image to be printed from a rubber typeface below the top plate. Since it is not known how many Morris' were actually manufactured, it is not believed to have been many. Few have actually survived and only a handful is known to exist. Serial number: none. (c.1887) Availability: Rare. Value: $10,000.

Munson

Two New Yorkers, James Eugene Munson and Samuel John Siefreid, teamed up to invent the Munson Typewriter. Patent date: September 17, 1889. The Munson typewriter had a metal sleeve-like tube for the typing element. It was removable and interchangeable, which was ideal for foreign language applications. It had a three row universal keyboard and was portable at only 16 pounds.

After the first model was released in 1890, an improved Model 2 followed. The Model 2 had a case added to cover the exposed parts of the machine. In 1897, Edgar A. Hill of Chicago, Illinois, purchased the Munson Company and changed the name to Chicago. He also introduced a Munson Model 3, which was followed very quickly by an identical machine, the Chicago Model 3.

Finally, production was standardized on a product that utilized a new case as pictured in the Model 3. The first standard model had some elaborate stenciling on the front of the case but subsequent models had a plain pin stripe. The Chicago was manufactured under many names: Draper, Baltimore, Conover, Galesburg, Yale and Ohio.

Munson (Model 1). Manufactured by the Munson Typewriter Company of Chicago, Illinois. Platen length: 8.5". It has three rows and 30-key keyboard with a double shift mechanism. Serial number: 1983c. (c.1890) Availability: Rare. Value: $3,000.

Munson (Model 3). Manufactured by the Munson Typewriter Company of Chicago, Illinois. This model has three patent dates stenciled on its case: September 17, 1889, February 20, 1891, May 3, 1892. Serial number: 9,703. (c.1899) Availability: Rare. Value: $2,500.

M & W

M & W. Manufactured by Gundka Werke, G.m.b.H. of Germany. Refer to Gundka for more information. Serial number: 12,286. (c.1924) Availability: Common. Value: $150.

Namograph

Namograph. Manufactured by the Midwestern Tool Company of Chicago, Illinois. This device printed letters on pencils. Serial number: 4,416. (c.1920) Availability: Rare. Value: $600.

National

National (Model 1a). Manufactured by the National Typewriter Company of Philadelphia, Pennsylvania. Invented and patented by Henry Harmon Unz in 1889. The National was a curved keyboard and under stroke device. The carriage had to be lifted up to see the type. A second under stroke National, the Model 1b, and was the same machine with a few minor differences, i.e. a larger space bar . Platen length: 8.5", two row 27-key curved keyboard. (c.1889) Availability: Rare. Value: $3,500.

National (Model 1b). Manufactured by the National Typewriter Company of Philadelphia, Pennsylvania. See National (Model 1a) for more information. Serial number: 3,208. (c.1889) Availability: Rare. Value: $3,500.

National (Rex)

National II (Model 2). Manufactured by the Rex Typewriter Corporation of Fond du Lac, Wisconsin. Originally introduced in 1917. In 1918, the American Can Company manufactured a Model 3 and then, in 1920, a Model 5 was also produced. In about 1920, a version was also manufactured as the Portex. Serial number: 528. (c.1917) Availability: Scarce. Value: $250.

National II (Model 3). See National II (Model 2) for more information. Serial number: 13,382. (c.1918) Availability: Scarce. Value: $250.

National II (Model 5). See National II (Model 2) for more information. Serial number: 22,540. (c.1918) Availability: Common. Value: $175.

NEMA

NEMA . This machine was manufactured for use by the Swiss army as a Secret Code machine. It was the successor to the famous ciphering machine – the Enigma. Only 640 machines were ever manufactured. Serial number: 286. (c.1947) Availability: Rare. Value: $3,500.

New American

New American (Model 5). Manufactured by the New American Manufacturing Company of Chicago, Illinois. (c.1890) Availability: Rare. Value: $2,200.

New Century

New Century (Model 5 – Caligraph). Manufactured by the American Writing Machine Company of New York, New York. After the Caligraph Model 4 and the turn of the 20th century, the American Typewriter Company released a new machine appropriately called the New Century. Although it still had a double keyboard, the space bar was moved to the traditional position and the general size and appearance resembled the more common typewriters of the period. The New Century is reported to have been designed by Walter J. Barron. Only two models have surfaced to date. A Model 5 in 1900, and a Model 6 in 1903. Serial number: 8,772. (c.1900) Availability: Scarce. Value Each: $250.

New Century (Model 6 – Caligraph). See information under New Century (Model 5 - Caligraph). Serial number: 11,655. (c.1903) Availability: Scarce. Value: $250.

Nippon

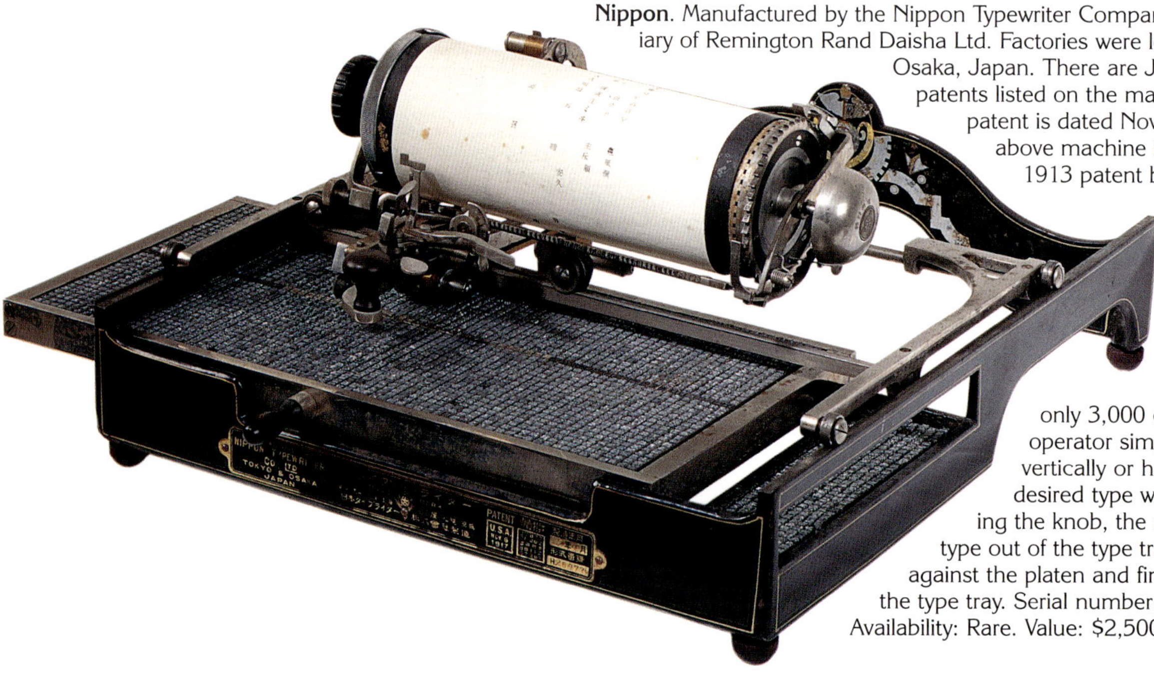

Nippon. Manufactured by the Nippon Typewriter Company of Japan, a subsidiary of Remington Rand Daisha Ltd. Factories were located in Tokyo and Osaka, Japan. There are Japanese and American patents listed on the machine. The American patent is dated November 6, 1917. The above machine is a later version of the 1913 patent by Kyoto Sugimoto. Although the Japanese language has in excess of 30,000 characters, the Nippon typewriter used a shorthand version called Katakana and contains only 3,000 characters. The operator simply moved the knob vertically or horizontally until the desired type was located. By depressing the knob, the mechanism lifted the type out of the type tray. Then it was pressed against the platen and finally, replaced back in the type tray. Serial number: 28,977. (c.1959) Availability: Rare. Value: $2,500.

Noiseless

The Noiseless typewriter was one of the great accomplishments of Wellington P. Kidder, who previously introduced the Wellington typewriter. Kidder collaborated with C. C. Colby on the development of this machine. Colby was a promoter and financier. In an effort to eliminate the annoying clatter of type banging against a platen in a generally noise-free business office, Kidder and Colby developed the pressure print theory, which resulted in the creation of the Noiseless. In 1896, they came up with a device based on their new principle for noise elimination. They theorized that they could replace the manner in which the hammer printed a character on paper by using what they called a "pressure or kissing impression." In 1904, after several years of work, a model was taken to the Parker Machine Company to make ready for production. This was not done overnight. In 1908, a factory was established in Middletown, Connecticut, and the name Noiseless Typewriter Company of Connecticut was born. However, no machines were produced for some time. In 1912, a major breakthrough occurred when the cam action was replaced by a new action called the momentum accumulator.

The principle of the momentum accumulator was simple. Using a weighted-toggle, the type gained momentum from the time the operator first struck the key, until it finally printed on the paper. The result was not only thequiet operation of the typewriter, but allowed for a much lighter touch. An operator no longer had to strike the keys, only tap them gently, with a kind of staccato touch. Still, there were no machines in production. It was not until 1915, that approximately 500 typewriters were finally produced and ready for sale.

A 1924 business publication lists the year as well as the serial numbers of those machines produced that year and is duplicated below:

Model	Year	Serial Numbers
	1915	500 (not listed in publication)
	1916	11,000
4	1917	13,000 (4 began with s/n 12,300)
	1918	17,000
	1919	24,000
	1920	35,000
	1921	51,000
5	1922	58,000 (5 began with s/n 60,000)
	1923	68,000

In 1921, a portable Noiseless typewriter was released. In 1923, the Model 5 was released which had 90 characters on a three-row keyboard. In 1924, Remington purchased the Noiseless Typewriter Company. The Noiseless was redesigned and introduced with a four-bank keyboard. It was called the Remington Model 6. A Remington advertisement of the period shows a Remington Noiseless Model 5 with a three-bank keyboard. Not many were made. A later advertisement shows the Remington Noiseless Model 6 with a four-row keyboard. The Model 6 was more common. The Noiseless was very popular in executive offices, hospitals, law offices and wherever a quiet atmosphere was desirable. It was common knowledge at Remington Rand that during the depression of the early 1930s, the Noiseless Typewriter was responsible for Remington's survival. The long career of the Noiseless typewriter extended well into the 1960s.

Noiseless (Model 1). Manufactured by the Noiseless Typewriter Company of Middletown, Connecticut. Serial number: 1-7,290. (c.1908) Availability: Rare. Value: $600.

Noiseless (Portable). Manufactured by the Noiseless Typewriter Company of Middletown, Connecticut. (See Noiseless (Model 1)) for more information. Serial Number: 2,245. (c.1921) Availability: Scarce. Value: $800.

North

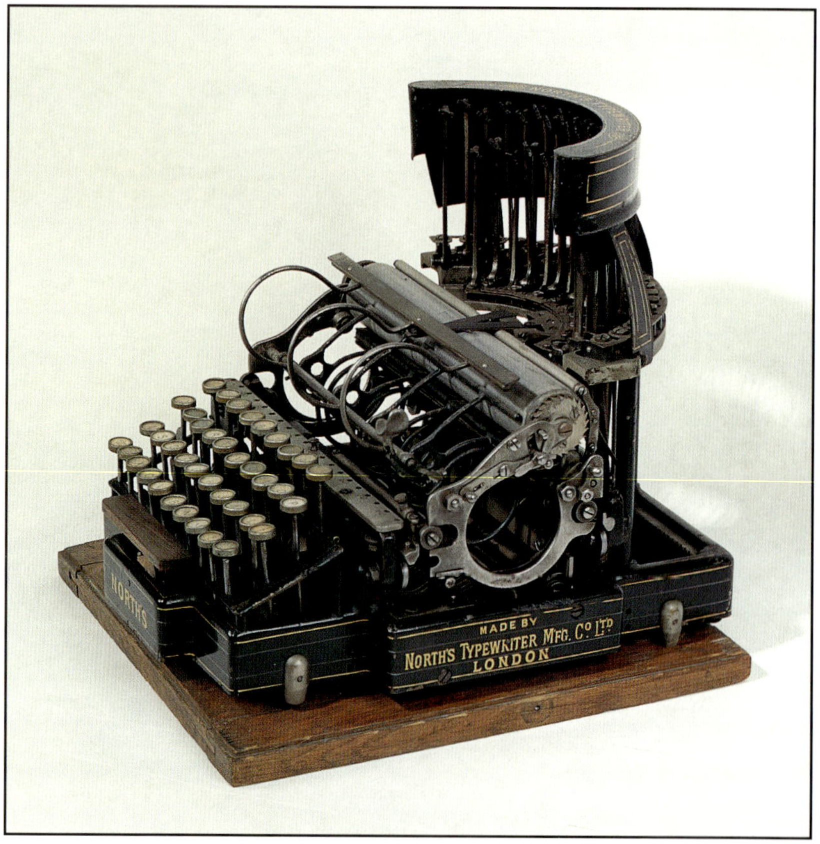

North. Manufactured by the North Typewriter Manufacturing Company, LTD., of London, England. Morgan Dunne and George B. Cooper invented the North in 1890. It is reported that the North got its name from its financial backer, Lord North. The North lasted only a short period of time after the death of Lord North. The type bars were arranged in two semicircles above and behind the platen. It was therefore, a down stroke machine. Serial number: 1,094. (c.1921) Availability: Rare. Value: $10,000.

Odell

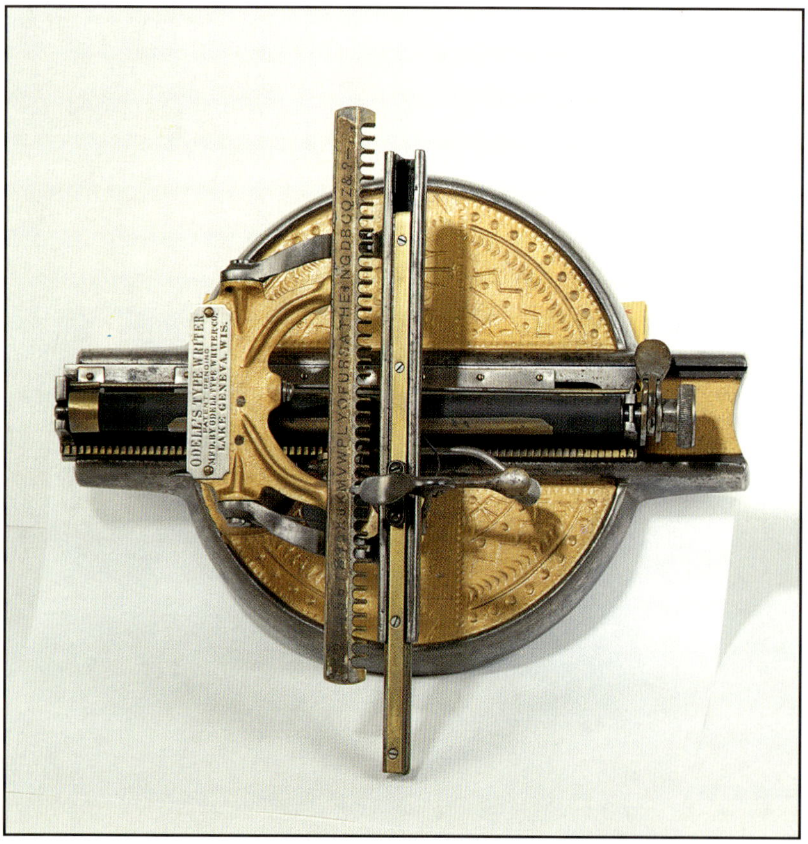

Odell (Model 1b). Manufactured by the Odell Typewriter Company of Chicago, Illinois. Invented and patented by L. J. Odell in 1889. Weighing only eight pounds, this indicator typewriter originally sold for $20. The price was its greatest feature. The type was arranged along a lineal shuttle and the operator had to use both hands to type. The left hand was used to add spaces and the right hand to depress the indicator key. Several models have surfaced: the Model 1, Model 1b, Model 2, Model 4 and Model 5. Serial number: None. (c.1889) Availability: Rare. Value: $2,000.

Odell (Model 2). Manufactured by the Odell Typewriter Company of Chicago, Illinois. See Odell (Model 1b) for more details. Serial number: 30,415. (c.1891) Availability: Scarce. Value: $800.

Odell (Model 4). Manufactured by Farquhar & Albrecht of Chicago, Illinois. (c.1894) Availability: Scarce. Value: $1,200.

Oliver

The Oliver Typewriter Company

Oliver (Model 2). Manufactured by the Oliver Typewriter Company of Chicago, Illinois. Platen length: 9" with a three-bank keyboard housing 28 keys. Serial number: 19,898. (c.1896) Availability: Scarce. Value: $500.

In 1888, Thomas Oliver, an Iowa minister, envisioned a need for a new kind of typewriter. A typewriter with concave type and type bars divided evenly on each side of the machine. The effect was an oblique down stroke method of typing. In 1892, Oliver was successful in securing his first patent. However, it was not until 1894 that Oliver sold his first machine to another minister. It was the Model 1 and the serial number was "3." That minister kept the machine for ten years before trading it in for a later model Oliver.

The Oliver Typewriter Company was incorporated in 1895. Perhaps the best analogy that testifies to the rapid growth of the company, was the fact that they outgrew their office space six times in the following ten years. Finally, in 1907 they built their own building.

Notable Oliver patents were are follows:

Model	**Patent Date**
Oliver Model 1	Patent Date: 1891
Oliver Model 2	Patent Date: October, 1894
Oliver Model 3	Patent Date: March, 1898
Oliver Model 5	Patent Date: May, 1907
Oliver Model 7	Patent Date: January, 1915
Oliver Model 9	Patent Date: March, 1916
Oliver Model 11	Patent Date: July, 1922

By some accounts, the Oliver Typewriter Company produced over 1,000,000 machines from 1894 to 1928 when they finally closed their doors. Therefore, because of the quantity manufactured and the indisputable durability, one is still able to find Oliver Typewriters in antique shops and flea markets.

Oliver (Model 2 – Nickel Plated). Manufactured by the Oliver Typewriter Company of Chicago, Illinois. Patent date: March 1, 1898. Serial number: 32,037. (c.1894) Availability: Scarce. Value: $700.

Oliver (Model 3). Manufactured by the Oliver Typewriter Company of Chicago, Illinois. Platen length: 8.875", a three-bank keyboard with 28 keys. Serial number: 152,182. (c.1898) Availability: Common. Value: $125.

Oliver (Model 5). Manufactured by the Oliver Typewriter Company of Chicago, Illinois. Platen length: 8.875", a three-bank keyboard with 28 keys. Patent date: August 24, 1909. Serial number: 385,314. (c.1907) Availability: Common. Value: $75.

Oliver (Model 7). Manufactured by the Oliver Typewriter Company of Chicago, Illinois. Patent date: June 10, 1913. Serial number: 563,510. (c.1915) Availability: Common. Value: $75.

Oliver (Model 9). Manufactured by the Oliver Typewriter Company of Chicago, Illinois. Patent date: June 10, 1913. Serial number: 644,787. (c.1916) This machine was found with two interchangeable carriages. The platen length of the carriage pictured on the machine is 8.75." The platen length of the extra carriage is 17.25." Both carriages have the same serial number. (NOTE: the Model 9 is common but since both carriages have the same serial number, the value is higher.) Availability: Scarce. Value: $300.

Oliver (Model 10). Manufactured by the Oliver Typewriter Company of Chicago, Illinois. Serial number: 100,732. (c.1922) Availability: Scarce. Value: $300.

Oliver (Model 11). Manufactured by the Oliver Typewriter Company of Chicago, Illinois. This was the first model that appeared without the side rabbit ears. Patent date: May 8, 1917. Serial number: 1,007,321. Availability: Scarce. Value: $300.

Oliver (Portable). Manufactured by the Oliver Typewriter Company. This machine was manufactured in Italy. Serial number: 66,701. (c.1928) Availability: Scarce. Value: $300.

Olivetti

Camillo Olivetti produced his first typewriter in 1910. That model was called the M1. While many consider Olivetti as the first successful Italian typewriter, it was not the first Italian machine. That distinction belongs to Guiseppi Ravissa's invention in about 1853. Olivetti's second machine was the Model M-20 that was introduced in 1920. Ten years later, the Model M-40 was introduced. Olivetti manufactured its first portable model the Model MP1 in 1932. In 1959 Olivetti took over the Underwood Company, and greatly expanded its American customer base.

Olivetti (Model – M1). Manufactured by the Olivetti Typewriter Company of Ivrea, Italy. This model was Italy's first commercially successful standard keyboard typewriter. (c.1910) Availability: Rare. Value: $2,550.

Olivetti (Model - M20). Manufactured by the Olivetti Typewriter Company of Ivrea, Italy. Platen length: 10.75" with 43 keys and 8-key decimal tabulation. Serial number: 54,466. (c.1920) Availability: Scarce. Value: $400.

Olivetti (Model – ECO - Portable). Manufactured by the Olivetti Typewriter Company of Ivrea, Italy. Platen length: 9.25" with 43 keys. Serial number: 321. (c.1932) Availability: Scarce. Value: $225.

Olivetti Portable (Model - Valentine). Made in Olivetti Plant in Barcelona, Spain. This is a very collectible portable, in spite of the fact, that is relatively of recent vintage. It was designed by Ettore Sottass of Italy. Platen length: 9.75" with 43 keys. Serial number: 5,247,320. Availability: Common. Value: $250.

Olivetti Portable Electric (Model - Lettera 36). Made in the German Democratic Republic. Serial number: 0,806,226. Platen length: 9.75", 43-key keyboard. Availability: Scarce. Value: $150.

Olivetti (Model – 201/221 – Electronic). Manufactured by the Olivetti Typewriter Company. This was Olivetti's first electronic typewriter. It was not the first electronic typewriter, however, as that distinction belongs to Qyx which was manufactured by the Exxon Oil Company in 1977. It is considered by some to the first commercially successful electronic typewriter. Serial number: 3,507,427. (c.1978) Availability: Common. Value: $150.

Olympia

The roots of the Olympia Typewriter Company date back to 1903, when the Mignon typewriter was their first product. The Mignon was a very popular indicator type machine. The parent company was the AEG (Allgemeine Elektrizitats-Gesellshaft). The Olympia name was first used in 1930 and eventually became their legal, worldwide title. Olympia became one of the great typewriter manufacturers of post- war Germany with global distribution.

Olympia (Model 8). Manufactured by Olympia Typewriter Company of Welmshaven, Germany. Platen length: 9.5", 44-key keyboard. Serial number: 265,478. (c.1932) Availability: Scarce. Value: $200.

Olympia (Model - Simplex [S]). Platen length: 9.5", 44-key keyboard. Serial number: 0,157,617. (c.1934) Availability: Common. Value: $150.

Olympia Robust (Model - WW II). Manufactured during World War II and used in Nazi headquarters. An American officer, who later become a State Senator from Maryland, took it as a souvenir. The carrying case is made to withstand the rigors of a battlefield and is painted a "gun-metal gray." Serial number: 413,919. (c.1942) Availability: Scarce. Value: $400.

Olympia Portable (Model - SGE 20). Platen length: 9.50", 44-key keyboard. Serial number: 72,256. Availability: Common. Value: $75.

Orga

Orga (Model 9). Manufactured by Bing Works of Nuremberg, Germany. The Orga was first introduced in 1922 and has 44 keys with a removable carriage for ease of cleaning and the ability to be interchanged with a longer carriage for statistical work. Serial number: 4,714. (c.1922) Availability: Common. Value: $125.

Orga (Model 10). Manufactured by Bing Works of Nuremberg, Germany. (See Orga (Model 9) for more information.) Serial number: 244,274. (c.1924) Availability: Common. Value: $150.

Patria

Patria (Model – Portable). Manufactured in Switzerland. Serial number: 17,954. (c.1937) Availability: Common. Value: $125.

Pearl

Pearl. Manufactured by the Pearl Typewriter Company of New York. Patented on May 12, 1891 by Theodore W. Searing. A circular index typewriter with characters printed around two thirds of the wheel. A pointer is used to select any one of 48 characters. Platen length: 8.75". Patent date: May 12, 1891. (c.1891) Availability: Rare. Value: $4,000.

Peerless

Peerless. Manufactured by the Peerless Typewriter Company of Ithaca, New York. It greatly resembled the Smith-Premier with a key for each character (76 in total). It was not on the market long and therefore you will not run into one very often. The inventors were Charles M. Clinton and James McNamara. Serial number: 3,726. (c.1891) Availability: Rare. Value: $1,000.

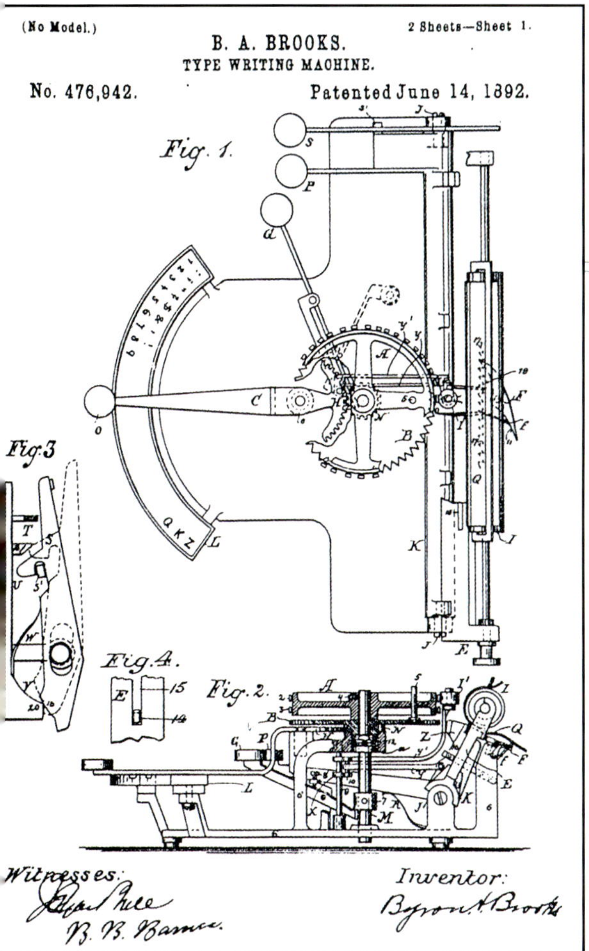

Peoples

Peoples (Model – Ink Roller). Manufactured by the Garvin Machine Company of New York, New York. On June 14, 1892, patent number 476,942 was issued to Byron Alden Brooks for an index typewriter that was to be inked using an ink roller. This feature distinguishes the first model from the second. A regular ribbon inked the second model. The second model also appeared under the name Champion. Serial number: 5,497. (c.1896) Availability: Scarce. Value: $800.

Perkeo

Perkeo (Model – Folding). Manufactured by Clemens Miller G.m.b.H. of Dresden, Germany. Almost identical to the Standard Folding Typewriter. Serial number: 20,014. (c.1912) Availability: Common. Value: $250.

Perkins

Perkins BrailleWriter (Model 1). Manufactured by the Perkins Institution of Watertown, Massachusetts. Serial number: 2. (Note: a very early serial number.) (c.1900) Availability: Scarce. Value: $600.

Perkins (Model - Easy Braille). Designed by David Abraham and manufactured and distributed by the Howe Memorial Press, Perkins School for the Blind in Watertown, Massachusetts. Serial number: E 4,346. (c.1972) Availability: Common. Value: $300.

Petite

Petite. Manufactured in Nottingham, England. (c.1960) Availability: Common. Value: $25.

Picht

Picht (Model - Braille). Manufactured by Van Herde & Wendy, Sebastiantr, 72 of Berlin, Germany. Invented in 1899 by Oscar Picht, who was an executive with the Institute for the Blind in Eberswalde, Germany. (c.1907) Availability: Common. Value: $200.

Pittsburg

Pittsburg Visible (Model 10). Manufactured by the Pittsburg Writing Company of Kittaning, Pennsylvania. A descendant of the Daugherty Typewriter, the Pittsburg Visible Model 10 also had a feature called the interchangeable type basket, which permitted optional keyboards, foreign languages and easier cleaning. Later models were redesigned to provide a smaller footprint while still maintaining the interchangeable type basket feature. Platen length: 9.75" with 43 keys. Serial number: 110,635. (c.1898) Availability: Scarce. Value: $800.

Pittsburg (Model 11). Manufactured by the Pittsburg Writing Company of Kittaning, Pennsylvania. Refer to Daugherty Visible and Pittsburg Visible (Model 10) for more information. Serial number: 35,689. (c.1908) Availability: Scarce. Value: $200.

Pittsburg (Model 12). Manufactured by the Pittsburg Writing Company of Kittaning, Pennsylvania. Refer to Daugherty Visible and Pittsburg Visible (Model 10) for more information. Platen length: 9.75" with 43 keys. Serial number: 35,537. (c.1914) Availability: Scarce. Value: $200.

Portex

Portex (Model 5). Manufactured by the Rex Typewriter Company of Fond du Lac, Wisconsin. A product created as a result of the National Typewriter Company being acquired by the Rex Typewriter Company. The name is a combination of Portable and Rex or Portex. It was a double shift, three-bank, front stroke typewriter that weighs only 9 pounds. Serial number: 19,020. (c.1922) Availability: Scarce. Value: $200.

Portex (Model 5 – Chrome). This particular Model 5 has a chrome finish and French keyboard. Platen length: 9.25" with 28 keys. Serial number: 22,222. Availability: Scarce. Value: $350.

Postal

Postal (Model 1). Manufactured by the Postal Typewriter Company of New York, New York. The first Postal was introduced in 1902 and manufactured at 45 Cliff Street in New York, New York. The Postal was invented by P. Quantrell and Franklin Judge. In 1904, manufacturing was moved to Norwalk, Connecticut. The frame is painted with a copper and black finish. A shield with label is pointing up and says "Postal, New York." Serial number: 1,953. (c.1902) Availability: Rare. Value: $900.

Postal (Model 3). Manufactured by the Postal Typewriter Company of New York, New York. The frame is painted black, shield is pointing down and says "Norwalk, Connecticut" on logo area. It also has a crude shift lock that is not on the Model 1. Serial number: 6,410. (c.1903) Availability: Scarce. Value: $700.

Postal (Model 5). Manufactured by the Postal Typewriter Company of New York, New York. All black frame with shield pointing down but "Model 5" is clearly imprinted on the bottom of the nickel-coated shield. It has a double shift on the left side of the keyboard to allow the operator to use all 84 characters housed on the 28 keys. Serial number: 25,226. (c.1904) Availability: Scarce. Value: $700.

Pyke-Glauser

Pyke-Glauser (BrailleWriter). National Institute for the Blind at 224-6-8 Great Portland Street, London, W.I. England. Serial number: 190. (c.1920) Availability: Scarce. Value: $200.

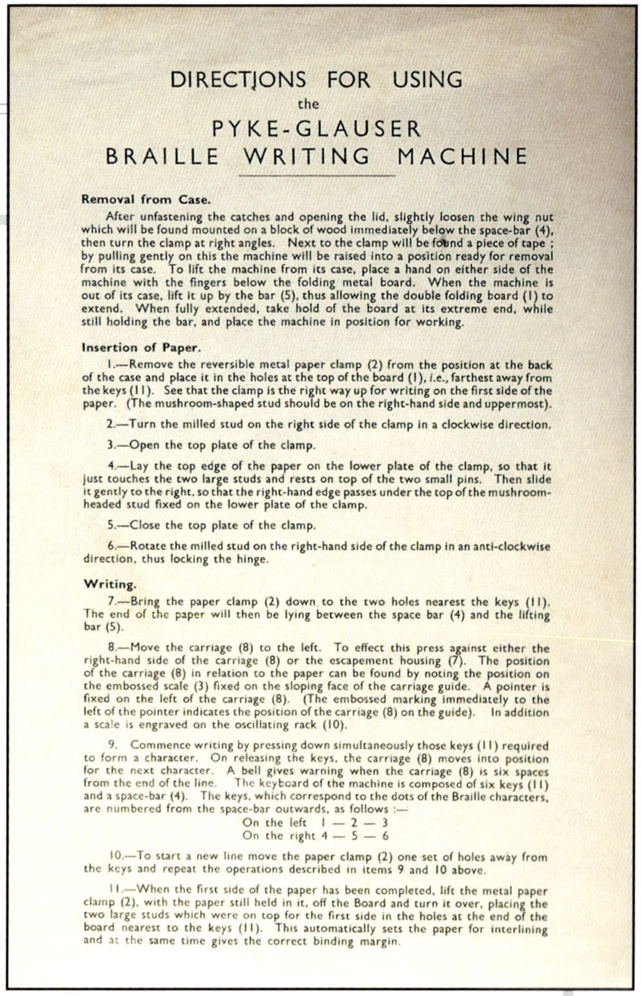

DIRECTJONS FOR USING
the
PYKE-GLAUSER
BRAILLE WRITING MACHINE

Removal from Case.

After unfastening the catches and opening the lid, slightly loosen the wing nut which will be found mounted on a block of wood immediately below the space-bar (4), then turn the clamp at right angles. Next to the clamp will be found a piece of tape ; by pulling gently on this the machine will be raised into a position ready for removal from its case. To lift the machine from its case, place a hand on either side of the machine with the fingers below the folding metal board. When the machine is out of its case, lift it up by the bar (5), thus allowing the double folding board (1) to extend. When fully extended, take hold of the board at its extreme end, while still holding the bar, and place the machine in position for working.

Insertion of Paper.

1.—Remove the reversible metal paper clamp (2) from the position at the back of the case and place it in the holes at the top of the board (1), *i.e.*, farthest away from the keys (11). See that the clamp is the right way up for writing on the first side of the paper. (The mushroom-shaped stud should be on the right-hand side and uppermost).

2.—Turn the milled stud on the right side of the clamp in a clockwise direction.

3.—Open the top plate of the clamp.

4.—Lay the top edge of the paper on the lower plate of the clamp, so that it just touches the two large studs and rests on top of the two small pins. Then slide it gently to the right, so that the right-hand edge passes under the top of the mushroom-headed stud fixed on the lower plate of the clamp.

5.—Close the top plate of the clamp.

6.—Rotate the milled stud on the right-hand side of the clamp in an anti-clockwise direction, thus locking the hinge.

Writing.

7.—Bring the paper clamp (2) down to the two holes nearest the keys (11). The end of the paper will then be lying between the space bar (4) and the lifting bar (5).

8.—Move the carriage (8) to the left. To effect this press against either the right-hand side of the carriage (8) or the escapement housing (7). The position of the carriage (8) in relation to the paper can be found by noting the position on the embossed scale (3) fixed on the sloping face of the carriage guide. A pointer is fixed on the left of the carriage (8). (The embossed marking immediately to the left of the pointer indicates the position of the carriage (8) on the guide). In addition a scale is engraved on the oscillating rack (10).

9. Commence writing by pressing down simultaneously those keys (11) required to form a character. On releasing the keys, the carriage (8) moves into position for the next character. A bell gives warning when the carriage (8) is six spaces from the end of the line. The keyboard of the machine is composed of six keys (11) and a space-bar (4). The keys, which correspond to the dots of the Braille characters, are numbered from the space-bar outwards, as follows :—

On the left 1 — 2 — 3
On the right 4 — 5 — 6

10.—To start a new line move the paper clamp (2) one set of holes away from the keys and repeat the operations described in items 9 and 10 above.

11.—When the first side of the paper has been completed, lift the metal paper clamp (2), with the paper still held in it, off the Board and turn it over, placing the two large studs which were on top for the first side in the holes at the end of the board nearest to the keys (11). This automatically sets the paper for interlining and at the same time gives the correct binding margin.

Post Office

Post Office Stamp Canceller. Used by West Chester, Pennsylvania Post Office to cancel stamps.

Qyx

Qyx (Model 2). Manufactured by the Exxon Corporation of Lyons, Pennsylvania. After acquiring the Vydec Word Processing Company, Exxon (an oil company) entered the office machine industry. In 1977, they announced their first product, the world's first electronic typewriter. The Qyx was large and bulky. It did not fit well on the conventional secretarial desk or workstation. In addition, Exxon/Qyx decided to market the product only on a direct basis, which proved to be very expensive. This did not allow the company to tap into an already established dealer customer base. As a result the product only survived a few years. Platen: 15" with 45 keys. Serial number: 49,671. (c.1977) Availability: Scarce. Value: $400.

Regina

Regina. Manufactured by Schilling and Kramer of Suhl, Thuringen. A front stroke, single shift machine that was originally manufactured as the Germanina. After some improvements it was released as the Regina. (c.1904) Serial number: 17,514. (c.1902) Availability: Common. Value: $150.

Reliance Visible

Reliance Visible. Manufactured by the Reliance Typewriter Company of Kittaning, Pennsylvania. After the Pittsburg Visible Typewriter Company was sold to the Reliance Visible Typewriter Company, the Reliance Visible was introduced. It was also marketed under the names Wall Street and Broadway Standard. Serial number: 57,640. (c.1915) Availability: Common. Value: $150.

Remington - *See Chapter II*

Rem-Sholes

Rem-Sholes (Model 4 – [or Rem (Fay) Sholes]). Manufactured by Remington Sholes Company. The sons of Christopher Latham Sholes and Eliaphalet Remington, Zalmon and Franklyn respectively, patented this machine in 1894 and introduced their first model in 1896. Remington (Wycoff, Seamans & Benedict) objected to their use of the Remington name and filed a lawsuit. After losing their case in the lower courts, the name was changed to simply the Fay Sholes. Charles N. Fay was the President at the time. The most laudable feature on the machine was the shift mechanism, which lifted the type basket instead of the carriage. This contributed to a lighter touch. The Rem-Sholes was an understroke machine which meant that the typing was not immediately visible. You must lift the carriage to see what had been typed. Some models, i.e., the one pictured here, were mistakenly called the Rem-Sho Bronze. They were actually not bronze. Instead they had a cast iron frame that was plated with oxidized copper. This machine was designed with beautiful columns and scrollwork that enhanced its appearance and contributes to its popularity among collectors today. A Supreme Court decision reversed the results of the previously mentioned lower court lawsuit and the original name (Rem-Sholes) could be used. There are at least ten models manufactured and their primary differences were in the number of keys on the keyboard. Serial number: 3,908. (c.1906) Availability: Scarce. Value: $1,500.

Reporter's Special

Reporter's Special. Manufactured by the Rex Typewriter Company and invented by D. C. Harris. It first appeared on the market in 1911 as the Harris Typewriter. Then the Harris company was sold to the Rex Typewriter Company and the Harris Typewriter was marketed under the name Reporter's Special. It was also sold as the Rex. Serial number: 20,282. (c.1916) Availability: Scarce. Value: $350.

Rex Visible

Rex Visible (Model 4). Manufactured by the Rex Typewriter Company. An all new typewriter introduced in 1916. D. C. Harris was the inventor. Serial number: 31,206. (c.1916) Availability: Common. Value: $75.

Rofa

Rofa (Curved Keyboard). Manufactured by Robert Fabig G.m.b.H. of Neurippen, Germany. Introduced in 1921, the Rofa has 30 keys with a double shift mechanism. Serial number: 19,593(a). Availability: Scarce. Value: $750.

Royal Barlock

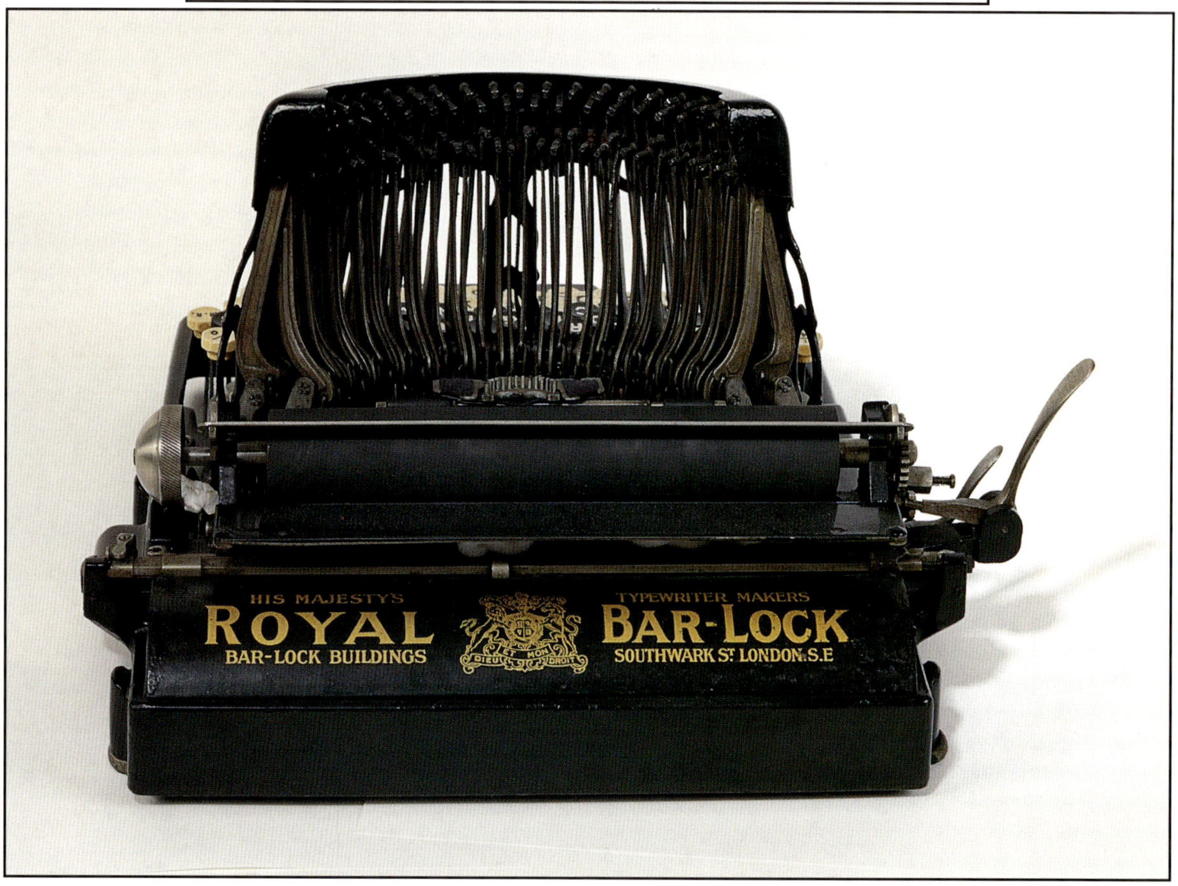

Royal Barlock. Manufactured by Bar-Lock, Southwark St., London, SE - England. Availability: Scarce. Value: $900.

Royal

The Royal typewriter first appeared on the market in 1906. It was invented by E. B. Hess and manufactured in Hartford, Connecticut. The focus of Hess's typewriter was in the area of touch. His principle innovation was described as the "reverse of accelerating toggle type bar movement." In practice, the operation would experience a light initial touch that would increase in speed as a result of the action. The end result was a typewriter that was perceived by the operators to have the lightest touch on the market and a company that would go on to become one of the great typewriter manufacturers of the world. Unfortunately, because so many were sold, very few have much value.

Royal Standard Statistics:

Year	Model, Serial number (if known) & other information
1906	Model 1 – Serial number 76,063
	Model 3 – long carriage version as per Beeching
	Model 6 – long carriage version as per Beeching
	Model 7 – long carriage version as per Beeching
	Model 8 – long carriage version as per Beeching
1911	Model 5 – Serial number 31,206 with two color ribbon, backspace, tabulator & paper finger.
1914	Model 10
1915	Model 10 – Serial number 174,74 with tabulator
1916	Model Q (Quiet Model) Royal Grand with concealed ribbon spools
1921	Model 10 (Royal Master Model)
	Model X – Serial number 736,659
1930	Model SX – Serial number 1,536,988
1935	Model H – Serial number x,xxx,xxx same as Model 10 but with different top plate.
1936	Model KHM
	Model KMM – Serial number 3,425.151
1947	Model KMM – Serial number 3,457,435 (Ten-Key Tabulator)
1949	Model KMG – Serial number 4,763.894
1950	Model RP – Serial number 4,812,141 (first Electric)
1952	Model HH – Serial number 6,074,670
1954	Model RE (second Electric)
1957	Model FP Manual
1960	Model HE Electric
1962	Model Electress
	Model Emperor – Deluxe Carbon Ribbon
1963	Model 101
1965	Model MC or Empress Model
1966	Model 440
	Models 660/550 – for Schools
1926	Model P – Serial number P59,888 in red and black
	Model P – Serial number P43,370 in green
	Model Junior – Serial number J50,675
1933	Model Signet Portable
1939	Model Quiet deLuxe – Serial number 1,903,707 (gold & Quiet deLuxe in Red) with Magic Margin.

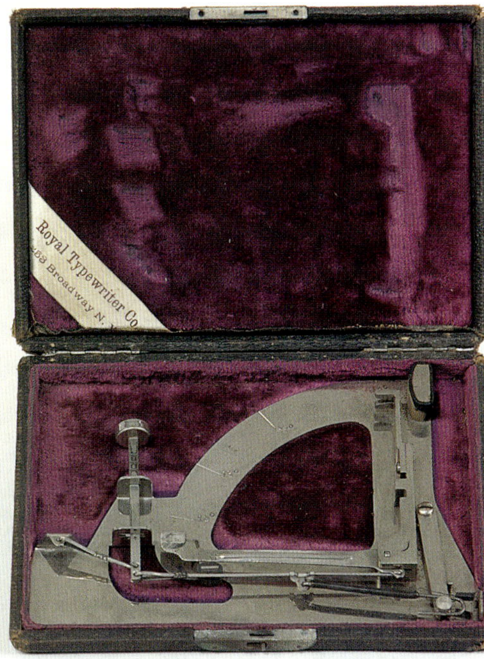

Royal (Touch Mechanism). A salesman's sample of the Royal type action which was used to demonstrate the Royal light touch that was preferred by operators.

Royal (Model 1). Manufactured by the Royal Typewriter Company of Hartford, Connecticut. Serial number: 76,063. (c.1906) Availability: Scarce. Value: $300.

Royal (Model 5). Manufactured by the Royal Typewriter Company of Hartford, Connecticut. Serial number: 31,206. (c.1911) Availability: Common. Value: $125.

Royal (Model 10). Manufactured by the Royal Typewriter Company of Hartford, Connecticut. Serial number: 174,775. (c.1914) Availability: Common. Value: $75.

Royal (Model - X). Manufactured by the Royal Typewriter Company of Hartford, Connecticut. Serial number: 736,659. (c.1922) Availability: Common. Value: $75.

Royal (Model - SX). Manufactured by the Royal Typewriter Company of Hartford, Connecticut. This was a 5-key tabulator model. Serial number: 1,536,988. (c.1930) Availability: Common. Value: $125.

Royal (Model - KHM). Manufactured by the Royal Typewriter Company of Hartford, Connecticut. Serial number: 2,080,298. (c.1936) Availability: Common. Value: $125.

Royal (Model - KMM w/Magic Margin). Manufactured by the Royal Typewriter Company of Hartford, Connecticut. Serial number: 3,435,151. (c.1947) Availability: Common. Value: $50.

Royal (Model - KMM w/TKT). Manufactured by the Royal Typewriter Company of Hartford, Connecticut. This was a ten-key decimal tabulator model. Serial number: 3,457,435. (c.1947) Availability: Common. Value: $175.

Royal (Model - KMG). Manufactured by the Royal Typewriter Company of Hartford, Connecticut. Serial number: 4,763,894. (c.1949) Availability: Common. Value: $50.

Royal (Model - HH). Early 1950's model that was introduced with a new design and a palm tabulator. Platen length: 10.5" with 43 keys. Serial number: 6,074,670. Availability: Common. Value: $50.

Royal (Model - RP). Manufactured by the Royal Typewriter Company of Hartford, Connecticut. First electric model. Serial number: 4,812,141. (c.1950) Availability: Scarce. Value: $200.

Royal (Model - Junior). Manufactured by the Royal Typewriter Company of Hartford, Connecticut. Serial number: J50,675. (c.1936) Availability: Scarce. Value: $125.

Royal (Model – Signet). Manufactured by the Royal Typewriter Company of Hartford, Connecticut. Only manufactured for two years. Serial number: E1,823. (c.1932-c.1933) Availability: Scarce. Value: $200.

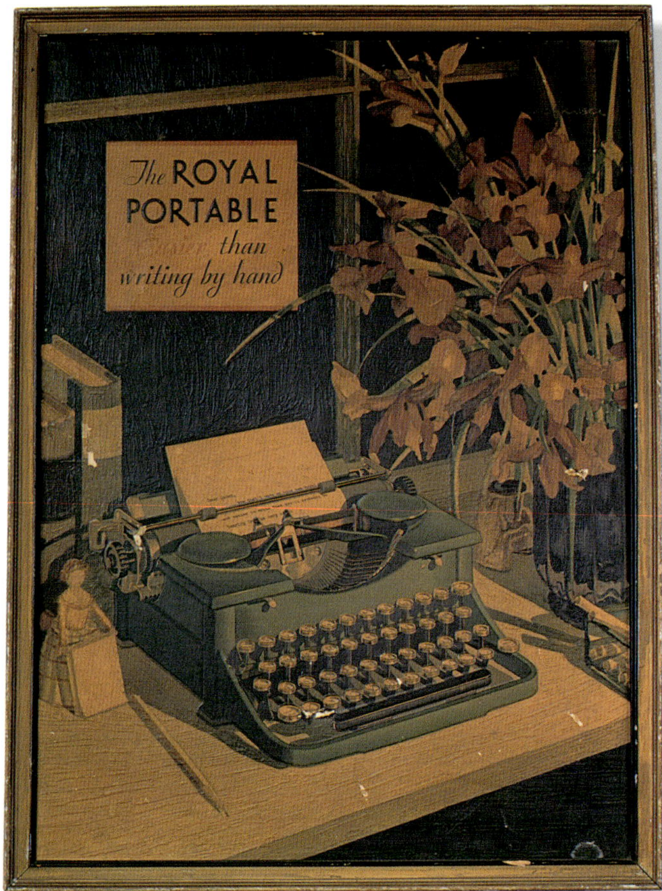
The ROYAL
PORTABLE
than
writing by hand

Bring it over!
We'll write it on my
ROYAL
PORTABLE

Royal (Model - P [Green] Portable). Manufactured by the Royal Typewriter Company of Hartford, Connecticut. Serial number: P43,370. (c.1926) Availability: Common. Value: $50.

Royal (Model P – [Red & Black] Portable). Manufactured by the Royal Typewriter Company of Hartford, Connecticut. Serial number: P59,888. (c.1926) Availability: Common. Value: $50.

Royal (Model – Quiet deLuxe Portable [Red]). Manufactured by the Royal Typewriter Company of Hartford, Connecticut. Platen length: 9.25" with 44 keys. Serial number: AP3,304,394. (c.1939) Availability: Common. Value: $125.

Royal Portable. With built in radio and speaker

Royal (Model – Quiet deLuxe Portable [Gold Plated]). Manufactured by the Royal Typewriter Company of Hartford, Connecticut. This was a special model released by Royal to celebrate their 50th Anniversary. Serial number: 1,903,707. (c.1956) Availability: Rare. Value: $1,000.

Royal (Model - Quiet DeLuxe). This was the 1950 version of the Quiet DeLuxe Portable. It had the same brown krinkle finish as the HH model Royal. Serial number: AG2,383,752. Availability: Common. Value: $25.

Royal (Model – 4 millionth Typewriter [Gold Plated]). The Royal Typewriter Company presented their 4 millionth typewriter to NOMDA [now BTA] in 1954. Serial number: 4,000,000. Availability: One of a kind. Value: Unknown.

Salter

Salter (Model - Improved 5). Manufactured by an English manufacturer of scales, George Salter and Company. Patented by James S. Foley and John H. Birch in 1892. Serial number: 2,039. Availability: Rare. Value: $5,000.

Salter (Model 6). Manufactured by George Salter & Company of West Brunswick, England. The inventor of the Salter was James S. Foley of London, England. The type and type bars were arranged in a semicircular fashion on the front of the platen, similar to the Bar-Lock. Initially, the early model used an inkpad but subsequent models were designed and equipped to use a ribbon mechanism. Platen length: 9" with 28 keys and a double shift mechanism. Serial number: 4,663. (c.1900) Availability: Rare. Value: $2,000.

Salter (Model 7). Platen length: 9" with 28 keys and a double shift mechanism . Serial number: 13,289. (c.1902) Availability: Rare. Value: $1,600.

Sampo

Sampo. Manufactured by Husqvarna Vapenfabriks A.B. of Sweden. Reported to be the first Swedish typewriter. Only 500 were manufactured. The author considers this to be a very desirable machine. It was developed and introduced by Dr. Raphael Herzberg. Today, Husqravna is well known for their line of electronic embroidery sewing machines. Serial number: 1,287. (c.1894) Availability: Rare. Value: $4,000.

Secor

Secor (Model 1). Manufactured by the Secor Typewriter Company of Derby, Connecticut. The inventor was Jerome B. Secor. He is reported to have worked on his invention for ten years. He began in 1901 and actually introduced the Model 1 in 1911. It is not surprising that it took Secor so long because he developed a typewriter with an easily removable escapement. Secor's removable escapement brings to mind the earlier Daugherty and Pittsburg, which has easily removable type baskets and keyboards. Both inventors were obviously looking for ways to disassemble and reassemble their machines to facilitate cleaning and maintenance, and also possibly to allow for optional type styles. In later years, Remington would design a standard typewriter called the Superwriter. After releasing 6 to 8 screws, the operator could literally unfold the typewriter. Remington called the feature, Fold-A-Matic construction. The typewriter could be easily refolded by simply lifting it from the rear. A nice feature, but, like the Daugherty and Secor, the benefit was mainly to maintenance personnel and not necessarily an attractive feature for the operator. About 7,000 Secors were manufactured. Manufacturing costs were exceptionally high and the Secor was withdrawn from the market in 1916. Platen length: 9.0625" and 42 keys. Serial number: B 4,246. (c.1911) Availability: Rare. Value: $700.

Secor (Model 2). Platen length: 9.375" with 42 keys. (c.1914) Availability: Scarce. Value: $500.

Senta

Senta. Manufactured by Frister & Rossmann of Berlin, Germany. (c.1914) Serial number: 19,355. Availability: Scarce. Value: $400.

Shimer

Shimer. The inventor of this machine was Elmer S. Shimer of Milton, Pennsylvania. John Lewis of Albuquerque, New Mexico, recently found an old Milton Machine Works catalog showing the Shimer in the centerfold. There are those who believe that the Shimer was never actually released for market; but the Shimer serial number would indicate otherwise. Only a few are known to exist. Serial number: 1,024. (c.1891) Availability: Rare. Value: $10,000.

Sholes & Glidden

While Christopher Latham Sholes was not the first person to invent the typewriter, he is generally regarded as the father of the typewriter. Most will agree that the Sholes & Glidden was the first practical typewriter that was commercially successful; it lasted for over 100 years. It was practical because Sholes and his colleagues worked together to produce a typewriter that was truly faster than the pen. Sholes initial idea was to link a telegrapher's key to a bar with a printer's type adhered to one end. By depressing the telegrapher's key, the type struck a flat surface, thereby causing an impression. This was the germ of the concept that was responsible for the three principle patents leading up to the Remington product eventually manufactured in 1873.

Unlike many inventors who worked alone, Sholes had a great deal of help from Carlos Glidden and Samuel W. Soule. All three men worked in a local Milwaukee machine shop, recruiting help from professional mechanical talents such as Mathias Schwalbach. But perhaps, the most important Sholes helper was James Densmore. Densmore was responsible for financing and marketing the new invention. It was Densmore who set up the meeting and presentation of the Sholes typewriter to the internationally known firearms manufacturer, E. Remington & Sons. With over 58 years of experience in manufacturing and marketing, Remington was positioned to give the Sholes typewriter its opportunity to revolutionize the American business office.

Although the typewriter presented to Remington was the result of years of work by Sholes and the last of 50 models, it was still crude in many respects. The Sholes typewriter was turned over to William K. Jenne, the Supervisor of the Sewing Machine production division at Remington. The first machines were released during the last few months of 1873. Only eight machines were manufactured that first year.

The first Sholes & Glidden machines were elaborately decorated to make them more user-friendly. These initial machines were also placed on sewing machine stands with the treadle connected to the carriage, so the carriage return was semiautomatic. It moved the carriage both horizontally and vertically. The first machines were all upper case, which immediately became an objection. Another objection was its selling price. One hundred twenty-five dollars ($125) was thought to be excessive, especially in a post-war recession period.

Sholes & Glidden (Model - Perfected). Manufactured by E. Remington & Sons of Elmira, New York. Serial number: 1,576. Availability: Rare. Value: $8,000.

There is ample evidence that Sholes and Densmore were the masterminds behind There is ample evidence that Sholes and Densmore were the masterminds behind the QWERTY keyboard arrangement. Both men were printers by trade. Perhaps the arrangement of the type in the printer's tray had some influence on their final keyboard design. But, the QWERTY keyboard we know today was the result of Sholes laboriously working and changing the positions of letters on the keys to minimize clashing and jamming the type bars at the printing point. Later, more scientific studies would be done to establish the best arrangement, but by that time, it was too late. The QWERTY keyboard had caught on and nothing could break the habit that had been ingrained in early tough typing classes.

The Sholes & Glidden basic concept would remain the standard for years to come. The vertical and horizontal platen and carriage; key tops linked to type bars and type; the QWERTY keyboard arrangement, all put together, made typing three times faster than the pen. It also allowed for the creation of carbon copies automatically and simultaneously without the strenuous use of a Copy Press.

Approximately 4,000 Sholes & Glidden machines were manufactured by Remington. There is a surprising number still around. They are desirable by collectors and museums around the world. To own a Sholes & Glidden is to own a piece of history. The Sholes & Glidden machines are considered to be national treasures and as such will command a high price for years to come.

Simplex

Simplex Toy (Model – Souvenir). Manufactured by the Simplex Typewriter Company of New York. This model was introduced in 1903 to commemorate the Centennial of the Louisiana Purchase (1803 to 1903). It has a picture of Thomas Jefferson and Napoleon Boneparte (the two people who negotiated the purchase) on the faceplate. Availability: Scarce. Value: $350.

Smith-Corona

Smith-Corona (Portable - Electric). Smith Corona's first portable electric typewriter. It had 44 keys and a 9.25" platen. The carriage return was done manually. There were two control discs located on the front of the machine. The one on the left was an impression control for carbon copies and the disc on the right was the on/off mechanism. Under the keyboard was a touch control. Serial number: 5TE-122,329. Availability: Scarce. Value: $125.

Smith, L. C. – *see L. C. Smith*

Smith Premier

Smith Premier (Model 1). Manufactured by the Smith Premier Typewriter Company of Syracuse, New York. The Smith Premier was patented by A. T. Brown in 1889 and introduced in 1890. It had a double keyboard and was an upstroke blindwriter. Ornate side panels that are nickel-plated distinguish the Model 1. Platen length: 8.625" with double keyboard. Serial number: 27,375. (c.1889) Availability: Scarce. Value: $300.

Smith Premier (Model 2). Platen length: 8.75" with double keyboard. Serial number: 72,823. (c.1896) Availability: Common. Value: $150.

Smith Premier (Model 3). Platen length: 13.5" Serial number: 5,595. (c.1896) Availability: Common. Value: $150.

Smith Premier (Model 4). Platen length: 8.75." Serial number: 56,825. (c.1896) Availability: Common. Value: $150.

Smith Premier (Model 10). Platen length: 9.375". Serial number: 76,359. (c.1908) Availability: Common. Value: $150.

Smith Premier (Model 5). Manufactured by L. C. Smith Gun Company of Syracuse, New York. Platen length: 11.5". Serial number: 12,941. (c.1901) Availability: Scarce. Value: $200.

Smith Premier (Model - Noiseless). Serial number: 112,731. (c.1928) Availability: Scarce. Value: $250.

Special

Special (Model 4 – AEG). Manufactured by Allgemeinen Electrigtato – Gesellshaft of Berlin, Germany. The Special – Model 4 was the same machine as the Mignon Model 4, but marketed under a different name. Refer to Mignon for more information. Serial number: 249,456. (c.1923) Availability: Scarce. Value: $300.

Standard Folding

Standard Folding (Models 1 & 2). Manufactured by the Rose Typewriter Company. The Standard Folding typewriter was invented from Frank S. Rose of New York, New York. Patent number 13,934 was reissued on June 22, 1915. After Rose died, the company changed its name to the Corona Typewriter Company. The Model 1 was a single shift machine and the Model 2 was a double shift. The Standard Folding was unique in that the platen and carriage folded over the keyboard making it even more compact for carrying, making it a popular portable, which the Corona Company sold very successfully for many years. As a result, the Corona is so common, that it has little value from a collector's standpoint. However, the Standard Folding model is scarce and therefore more collectible. Serial numbers: 3,687; 8,749 respectively. (c.1907) Availability: Scarce. Value: $250 each.

Staensby-Wayne

Staensby-Wayne (Model – Shorthand). Manufactured by the Alfred Wayne & Company of Handsworth, Birmingham in England. The inventor of this machine was Henry Staensby. Serial number: 502. (c.1899) Availability: Scarce. Value: $300.

Staensby-Wayne (Model - Braille Writer). Manufactured by the Alfred Wayne & Company of Handsworth, Birmingham in England. The inventor of this machine was Henry Staensby. Serial number: 3351/61. (c.1899) Availability: Scarce. Value: $300.

Stearns Visible

Stearns Visible. Manufactured by E. C. Stearns & Company of Syracuse, New York. A front stroke type bar machine with a universal keyboard of four rows of keys. As the model name indicates, type was completely visible to the operator. It has a unique tabulator lever on the right side of the front panel that can be moved to select eight positions and is reported to have been discontinued around 1920. Serial number: 1,577. (c.1900) Availability: Scarce. Value: $200.

Stenograph

Stenograph. Manufactured by the United States Stenograph Company of St. Louis, Missouri. Was originally invented by Miles Bartholomew in 1879, but not introduced until 1882. The keyboard consisted of nine keys and a space bar. A ticker tape, like roll of paper, was used as the media on which a coded result was printed. Serial number: 2,013. (c.1882) Availability: Rare. Value: $1,000.

Stenotype

Stenotype. Manufactured by the Universal Stenotype Company of Owensboro, Kentucky. A shorthand machine invented by Ward Ireland of Dallas, Texas. It produced a system of shorthand called stenotype, in which all silent letters were omitted. One or several keys could represent an entire word or even whole phrases and are written with one stroke. Serial number: 3,909. (c.1911) Availability: Common. Value: $75.

Sterling

Sterling (Model [a]). Manufactured by the Sterling Typewriter Company, 115 Broadway, New York, New York. Platen length: 9.5" with 31 keys and a double shift. It used a swinging shuttle type similar to the Hammond. (c.1910) Availability: Rare. Value: $4,500.

Sterling (Model [b]) - *See Fox Typewriter*

Stoewer Record

Stoewer Record. Manufactured by the Bernard Stoewer Actien-Gesellschaft Stettin of Grunhof, Germany. The first model appears in 1903 and was called appropriately the Model I. It was followed by the Models II, III and IV. All appeared before 1909 when the Stoewer Record was released. A portable, Stoewer Elite came out in 1912. Serial number: 134,481. (c.1903) Availability: Common. Value: $125.

Sun

Sun (Model 2). Manufactured by the Sun Typewriter Company of New York, New York. At least nine models of the Sun were released, but by far the most popular was the Model 2. The most impressive feature of the Model 2 was a principle of "direct inking." This not only produced a crisp, sharp impression, but also gave more legible carbon copies. Platen length: 9.875" with 27 keys. Serial number: 12,276. (c.1901) Availability: Scarce. Value: $400.

Surety

Surety (Model A). Manufactured by the American Typewriter Company. Serial number: 25,518. (c.1904) Availability: Scarce. Value: $400.

Taurus

Taurus (Model – Pocket). Manufactured by Torrani & Company of Milan, Italy. It looks like a pocket watch with typeface printed cylindrically around the dial or face. Selecting a character by rotating the dial and depressing a watch like plunger at the top causes an impression on paper tape (not unlike a ticker or teletype tape). It is probably considered to be the most desirable of collectible Italian typewriters. Few are known to exist. (c.1908) Availability: Rare. Value: $9,000.

Teletypewriter

Teletypewriter (Model 283A). Manufactured by Kleinschmidt Laboratories, Inc. of Deerfield, Illinois. A U. S. Army Signal Corps Model. Serial number: 3,506. (c.1950) Availability: Scarce. Value: $300.

Titania

Titania. Manufactured by the Titania Typewriter Company of Berlin, Germany. Platen length: 9.75" with 42 keys. Serial number: 7,401. (c.1910) Availability: Common. Value: $150.

Torpedo

Torpedo Portable. Manufactured by Weil-Werke, Germany. Platen length: 9.5" with 42 keys. Serial number: 675,938. (c.1953) Availability: Common. Value: $75.

Tower

Tower (Model - Chieftain II). Distributed by Sears Roebuck and Company. Platen length: 8.75". 42-key keyboard. Serial number: 3YT 156,881. Availability: Common. Value: $25.

Toyriter

Toyriter. Manufactured by the Wayne Toy Manufacturing Company of Dayton, Ohio. A well-made dial indicator typewriter, using inkpads and rubber type arranged around the circumference of the dial face. Availability: Rare. Value: $500.

Travis

Travis. Manufactured by the Travis Typewriter Company of Philadelphia, Pennsylvania. The Travis was invented by William H. Travis and patented April 7, 1896. It utilized a type wheel and a narrow ribbon that is mounted vertically on each side of the print wheel. The Travis was only marketed for a very short period of time and only a few examples are known to exist. (c.1905) Availability: Rare. Value: $4,000.

Triumph

Triumph (Model - Norm 6). Manufactured by Triumph-Werke of Nuremberg, Germany. A German four-bank, single shift offshoot of the Norica Typewriter. It was also available with a ten-key decimal tabulator. 1921 marked the introduction of the company's last model. The first manufacturer of the Triumph was the Deutsch-Triumph-Fakrradwerke, S. G., a bicycle manufacturer. Internationally it was sold under various names, i.e.: Mondiale (France); Edita (South America); Phoenix and Koh-I-Noor. The Triumph-Werke Company also manufactured portables. Platen length: 9-1/8" with 44 keys. Serial number: 430,096. (c.1910) Availability: Common. Value: $200.

Triumph (Model 2). Manufactured by Triumph Werke A.G., Nuremberg, Germany. Platen length: 9.25" with 44 keys. Serial number: 34,323. (c.1911) Availability: Common. Value: $125.

Triumph (Model – Durabel). Platen length: 9.5" with 44 keys. Serial number: 318,082. Availability: Common. Value: $125.

Typatune

Typatune Toy. Manufactured by the Electronic Corporation of America of New York. It is a toy typewriter on which a tune can be played just as its name implies. Patent number: 215,949. Availability: Scarce. Value: $200.

Typo

Typo. Manufactured by Francaise d'Armes et Cycles de Saint-Etienne of France. The Typo is the same machine as the British Imperial Model B, but was manufactured and marketed in France. Serial number: 9,794. (c.1908) Availability: Scarce. Value: $1,000.

Typograph

Typograph (Cash). Manufactured by the Typograph Company. Arthur W. Cash invented this flat bed, down stroke typewriter in 1897. It was originally produced as the Cash. It was designed to type on extra heavy paper or card stock. Few have survived and therefore it is desirable from a collector's point of view. Patented September 2, 1884. Availability: Rare. Value: $8,000.

Underwood

The Underwood typewriter was invented by Franz Xavier Wagner in 1892 and first manufactured by the Wagner Typewriter Company. In 1895, John T. Underwood purchased the Wagner patents and started the Underwood Typewriter Company. The Underwood family was well known in the industry as manufacturers of typewriter ribbons and carbon paper. One story that has circulated throughout the years is that Underwood was the supplier of all Remington typewriter ribbons and carbon paper for years. The time came when Remington decided to compete with Underwood in the production of these supplies. Remington cancelled their supply agreement with Underwood and began to manufacturer supplies themselves. Underwood was so furious that they decided to compete with Remington typewriters. Subsequently, Underwood purchased the Wagner Typewriter and began typewriter production. So, Remington was instrumental in creating their biggest competitor.

Underwood dominated the American typewriter industry for many years until the Italian firm of Olivetti purchased them in 1959.

Underwood (Model 1). Manufactured by the Wagner Typewriter Company in 1893. Wagner Typewriter Company became the Underwood Typewriter Company in 1895. The Underwood was one of the most popular typewriters of all times. A light touch, basket shift and visible type were among the most important reasons for its immense popularity. Serial number: 8,558. (c.1893) Availability: Rare. Value: $600.

Underwood (Model 3). See Underwood (Model 1) for more information. Serial number: 3,801,197. (c.1919) Availability: Scarce. Value: $150.

Underwood (Model 3 – TKT). See Underwood (Model 1) and Remington (Model 7 – TKT) for more information. Patent dates: August 26, 1919 (ten-key tabulator mechanism), August 14, 1921 (typewriter). Serial number: 514,030-12. Availability: Scarce. Value: $200.

Underwood (Model 4). See Underwood (Model 1) for more information. Serial number: 430,517. (c.1901) Availability: Common. Value: $175.

Underwood (Model 5). See Underwood (Model 1) for more information. Serial number: 1,780,020-5. (c.1901) Availability: Common. Value: $200.

Underwood (Model – Noiseless). Under license by Remington, Underwood produced the Noiseless typewriter. Platen length: 10.5" with 42 keys. Serial number: N4,821,575. (c.1936) Availability: Common. Value: $50.

Underwood (Model – 3 Bank Portable). See Underwood for more information. Platen length: 9.25" with 28 keys and a double shift mechanism. Serial number: 82,225. (c.1919) Availability: Common. Value: $50.

Underwood (Model – Noiseless Portable). Manufactured under license by Remington. Platen length: 9" with 42 keys. Serial number: 4,821,575. (c.1931) Availability: Common. Value: $50.

Underwood (Model – Portable [Universal - U. S. Navy]). Manufactured by Underwood/Elliott-Fisher. Platen length: 9" with 42 keys. Serial number: F1,426.674. Availability: Common. Value: $125.

Underwood Electric. Platen length: 14.5" with 42 keys. Serial number: 6,689.066. (c.1951) Availability: Common. Value: $150.

Underwood (Model AA - Quiet Tab Portable). Platen length: 9.5", 43-key keyboard. Serial number: 2,668,517. Availability: Common. Value: $75.

Underwood (Model 478 - Electric Portable). Platen length: 9.75" with 43 keys and electric carriage return. Serial number: 619,424. Availability: Common. Value: $25.

Underwood (Model - Leader Portable). Platen length: 9.625", 42-key keyboard. Serial number: CA 2,703,732. Availability: Common. Value: $50.

Urania

Urania Portable. Manufactured by Clemens Muller G.m.b.H. of Germany. Serial number: 118,450. (c.1909) Availability: Common. Value: $25.

Venus

Venus. Manufactured by Venus Werf Neugersdorf, Germany. It has 46 keys and a 9.5" platen. Serial number: 671. Availability: Scarce. Value: $800.

Victor

The Victor type-bar machine was invented by J. A. Hagerstrom and first patented on December 5, 1891. A unique feature of the Victor was the ribbon vibrator mechanism. Instead of the ribbon being fed horizontally, with black on the top position and red on the bottom, the ribbon was fed vertically, with black on the left and red on the right position. The Victor Adding Machine Company of Chicago, Illinois, purchased the patents and name of the Victor Typewriter Company and subsequently introduced a portable typewriter.

Victor (Model – Index). Manufactured by the Tilton Manufacturing Company of Boston, Massachusetts. This invention of Frank Taylor and J. A. White of Hartford, Connecticut, was patented in 1889. It was an index machine with a type sheet like element. A. I. Jacobs patented later improvements. Patent number: 409,289. In 1891, E. G. Young developed an electrified model of the Victor Index. Patent number: 449,923. Platen length: 7.75" Serial number: 1,571. (c.1889) Availability: Rare. Value: $3,000.

Victor (Model 3). Manufactured by the Victor Typewriter Company of New York, New York. First patent: December 5, 1891 and last patent dated November 3, 1909. Platen length: 9.25" with 42 keys and eight-key decimal tabulation. Serial number: 20,979. (c.1910) Availability: Scarce. Value: $250.

Victor (Model 10). Manufactured by the Victor Typewriter Company of New York, New York. Patented: September 9, 1919. Platen length: 10" with 42 keys. Serial number: 50,908. (c.1919) Availability: Common. Value: $75.

Victor (Model 3 – Portable). Manufactured by the Victor Adding Machine Company of Chicago, Illinois. Platen length: 9.5" with 42 keys. Serial number: 100,002. Availability: Scarce. Value: $300.

Virotype

Virotype. Manufactured by Machines a ecrire Virotype, 30 Rue de Rickelien, Paris, France. Designed in France by H. Viry in 1914. Its purpose was to be portable and convenient for use by military personnel in field combat situations. Serial number: 7,572. (c.1914) Availability: Scarce. Value: $800.

Vydec

Vydec. The first full-page CRT based word processing machine introduced in the early 1980s . The machine was composed of several different components.

Wellington

Another gifted American inventor, Wellington Parker Kidder of Jamaica Planes, Massachusetts, patented the Wellington in 1892. The Wellington was manufactured by the Williams Manufacturing Company of Plattsburg, New York, and was also manufactured in Montreal, Canada, as the Empire and in Germany as the Adler. This forward thrust machine made its first appearance in 1885. It had a three-bank keyboard and a nice typing touch or action, which is probably why some form of the forward thrust design was marketed until the end of 1920. A similar type action principle was introduced in 1912 as the Noiseless Typewriter.

Wellington. Manufactured by Williams Manufacturing Company of Plattsburg, New York. Platen length: 8.25" with 28 keys and a double shift. Serial number: 18,959. Availability: Common. Value: $200.

Wellington (Model 2). Platen length: 8.5" with 28 keys and double shift. Serial number: 5,461. Availability: Common. Value: $175.

Williams

Manufactured by the Williams Typewriter Company of Derby, Connecticut. The little state of Maine was not to be denied their place in typewriter history. John Newton Williams of that state invented this unique machine. The type bars were divided, with half in front of the platen and half behind. As a key was depressed, the type bar would arc toward the platen in a movement similar to the movement of a grasshopper. As a result, the Williams is often referred to as the grasshopper typewriter. The first model was a 28-key, curved keyboard – introduced in 1892. Collectors refer to it as the Model 1-Curved to distinguish it from the Model 1-Straight introduced in 1895. The Williams Company was placed in receivership in 1909. After a short period of reorganization, the company became the Secor Typewriter Company. In 1912, the Williams Model 2 was reintroduced as the Junior by Secor.

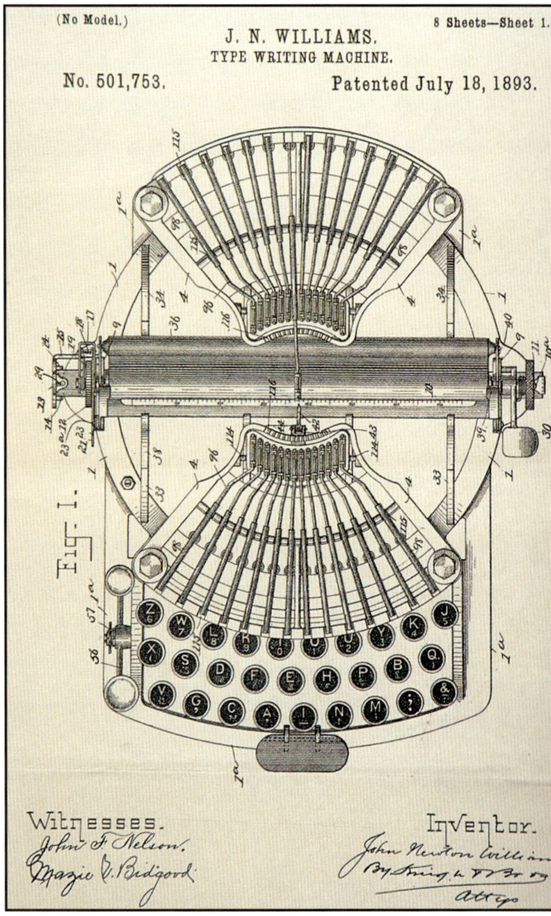

Williams (Model 1 – Curved). Manufactured by Brady Manufacturing Company of Brooklyn, New York. Serial number: 2,087. (c.1891) Availability: Rare. Value: $8,000.

Williams (Model 1 – Straight). Manufactured by the Williams Typewriter Company of Derby, Connecticut. Platen length: 7.75" with 28 keys. Serial number: 6,187. (c.1893) Availability: Rare. Value: $2,000.

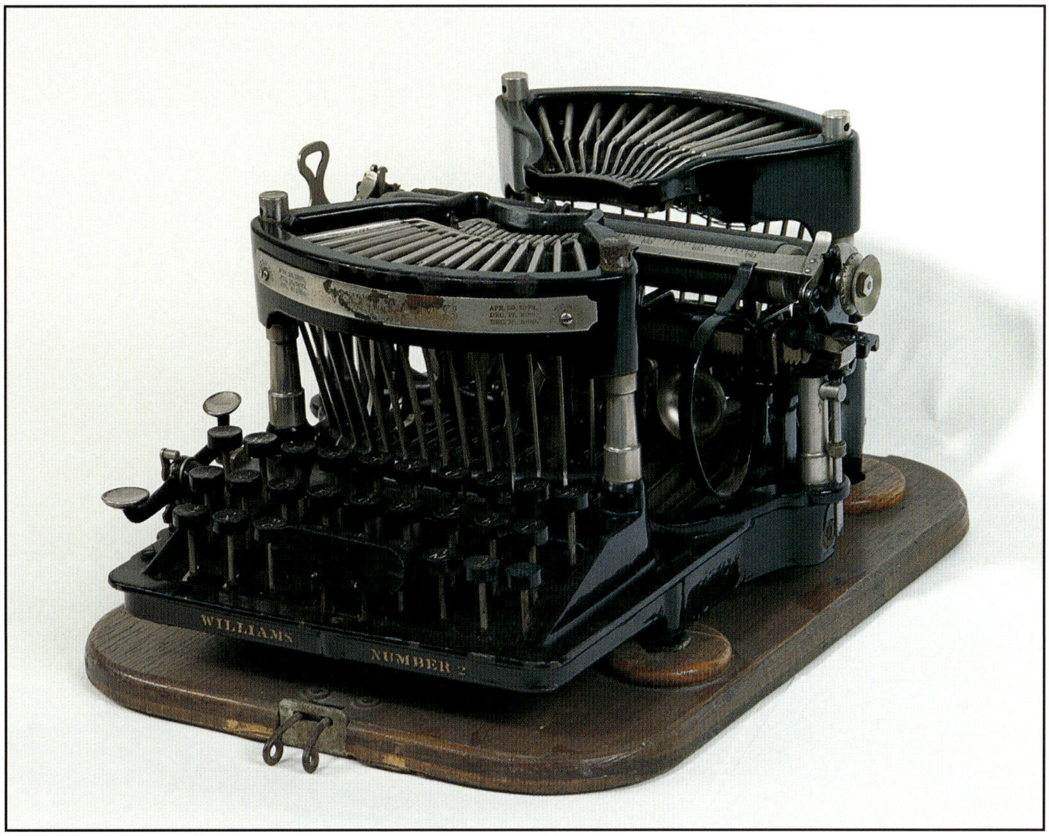

Williams (Model 2 - Straight). Platen length: 7.75" with 28 keys and a double shift mechanism. (c.1896) Serial number: 10,388. Availability: Scarce. Value: $800.

Williams (Model 4). Platen length: 9.375" with 42 keys. Serial number: 25,282. (c.1900) Availability: Scarce. Value: $300.

Williams (Model 6). Platen length: 9.125" with 42 keys. Serial number: 37,056. (c.1906) Availability: Scarce. Value: $300.

Woodstock

Woodstock (Model 5). Manufactured by the Woodstock Typewriter Company of Woodstock, Illinois. Two additional wide carriage machines were introduced in 1922, which were basically the Model 5. The Model 6 had a 14.5" carriage. In 1923 an 18.5" carriage Model 7 was released. Serial number: N370,450. Availability: Common. Value: $50.

Woodstock (Model – Electric). Manufactured by the Woodstock Typewriter Company of Woodstock, Illinois. Introduced in 1924. This was not a true electric typewriter. Only the carriage return was electric and not the keys themselves. Serial number: E -14 - 495,202 E. (c.1924) Availability: Scarce. Value: $300.

World

World (Model 1). Manufactured by the Typewriter Improvement Company of Boston, Massachusetts. Invented by John Becker of the same city. This was an index typewriter and was first manufactured by the World Typewriter Company of Maine. In later years it was manufactured by a bicycle company called the Pope Manufacturing Company, and, still later, by the Typewriter Improvement Company. (c.1886) Availability: Scarce. Value: $400 each.

Write Easy

Write Easy (Model – Index). Manufactured by Gundka Werk, G.m.b.H. of Germany. See Gundka. Serial number: 86,670. Availability: Scarce. Value: $200.

Yetman

Yetman. Manufactured by the Yetman Typewriter Transmitter Company of New York, New York. Charles E. Yetman of Washington, D.C., is credited with inventing this remarkable typewriter. The purpose of this invention was to solve the problem of needing a skilled telegraph operator to send and/or receive Morse code typed messages. This typewriter performed an internal conversion process to change the Morse code signal to readable English and vise versa. For its time, it was a great idea. It potentially turned the average typist station into a communicating typewriter with the ability to send and receive international messages that could be instantly read. Unfortunately, Yetman died within a year of the introduction of his machine and the company was sold to J. L. Schmidt of Philadelphia, Pennsylvania. After a short period of time, this machine was reintroduced as the Smith Visible. Some models of the Yetman were only transmitting machines and others were both transmitters and receivers, as is the author's machine. Serial number: 677. (c.1907) Availability: Rare. Value: $5,000.

Yost

Although the Yost Typewriter was not invented by George Washington Napoleon Yost, he was the driving force behind the marketing of the product. There were three individuals responsible for patenting the Yost in 1889. They were Alex Davidson, Andrew Steiger, and Jacob Febel. G. W. N. Yost did contribute some improvement to the machine as shown by his patent number 450,806 of April 21, 1894. Yost was one of the pioneers of the Sholes & Glidden Typewriter as well as the Caligraph, and therefore, brought a great deal of experience to the Yost Typewriter Company.

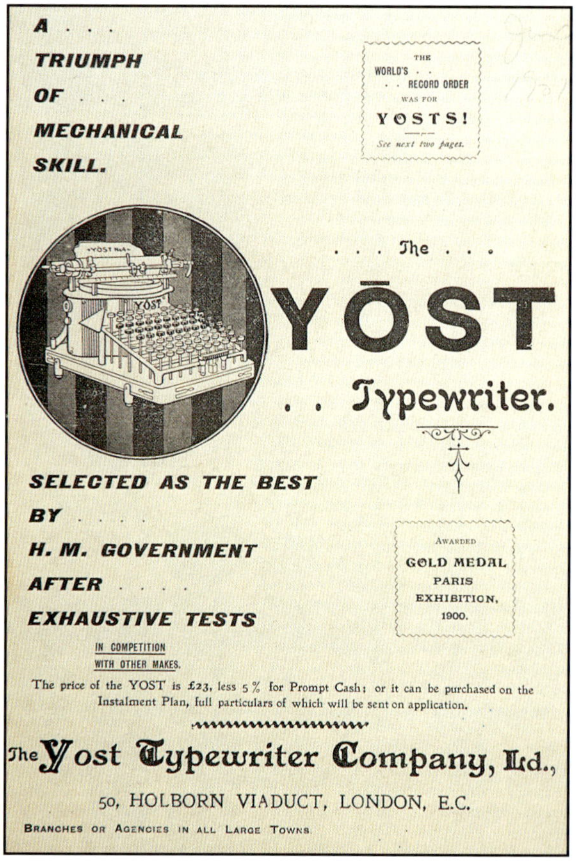

Yost Advertisement.

Yost (Model 1). Manufactured by the Yost Writing Machine Company of Bridgeport, Connecticut, at the Merritt Brothers Factory, Springfield, Massachusetts. The Model 1 Yost was a blind writer and double keyboard machine. It had an alignment system whereby the type traveled through a square metal device that insured perfect alignment. The Model 1 also had two medallions embedded in the two rear corners of the back frame. Serial number: 1,888. (c.1889) Availability: Rare. Value: $1,000.

Yost (Model 4). Called the New Yost. Platen length: 8.625" with double keyboard. Serial number: 29,970. (c.1895) Availability: Scarce. Value: $700.

Yost (Model 10). Although most Yost Model 10 typewriters were of the new four-row visible variety, there were a few of the Model 10s with double keyboards. The picture above is an example of the latter. Platen length: 8.875" with a double keyboard. Serial number: 73,761. (c.1902) Availability: Scarce. Value: $700.

Yost (Model 15). This was an example of the new design with a four-row keyboard. Platen length: 10.875" with 42 keys and a four-key tabulator. Serial number: 2,850,204. (c.1908) Availability: Scarce. Value: $500.

Young American

Young American. Manufactured by the American Typewriter Company of New York, New York, incorporated in 1893. See American Visible for more information. Serial number: 4,183. Availability: Rare. Value: $1,000.

Yu Ess

Yu Ess. Manufactured by the Yu Ess Manufacturing Company of New York, New York. Derived from the basic patents of the Mignon Typewriter Company of Germany. Although the Yu Ess was introduced in 1918, it was withdrawn from the market in 1919, so, not many are around today. (c.1918) Availability: Rare. Value: $800.

Chapter Six
THE LEGACY OF MECHANICAL TYPEWRITERS

When the Sholes & Glidden (Remington) typewriter made its first public appearance at the Philadelphia Centennial in 1876, it definitely did not receive the attention it deserved. The hit of the exhibit was Alexander Graham Bell's telephone. And, rightly so, because the telephone did change our lives enormously.

But, the typewriter offered something equally important, if not more so . It changed the lives of women forever. The typewriter created a new occupation. Millions of stenographic positions were created almost exclusively for women. Writing speeds escalated from 30 words per minute (wpm) longhand to as much as 136 wpm by typewriter.

Typing eliminated the mess of pen and ink. In addition, it allowed for multiple carbon copies that eliminated the need for the copy press.

The touch method of typing made writing easier, faster, and better than ever before.

The QWERTY keyboard originally developed by Christopher Latham Sholes and James Densmore dates back to circa 1872, but, it is still with us in the new "twenty-first century" as an integral part of our modern day computers and remains as the input method of choice.

The contributions of the typewriter are many. However, the overwhelming legacy received from the typewriter is, that it ushered women into the office work force. It tapped the other half of our intellectual base. It allowed women to become independent. It gave them the ladder on which they could climb to the top of the business world. Of course, the typewriter did not guarantee their success, but it created a path and opened a window of opportunity. The rest was up to them, and their track record is impressive.

The following represents a partial tribute to those wonderful women who used the typewriter to achieve their place in the world's work place.

Christopher Latham Sholes's daughter at the typewriter.

Christopher Latham Sholes's daughter rendering from the *Scientific American,* August 10, 1872.

Woodcut of a typist that appeared on the first typewriter brochure distributed at the Philadelphia, Pennsylvania 1876 Centennial Celebration.

Artist's rendering of a typist using a Beech typewriter as depicted in the *Scientific American,* December 18, 1896.

Early Fox typist, name unknown.

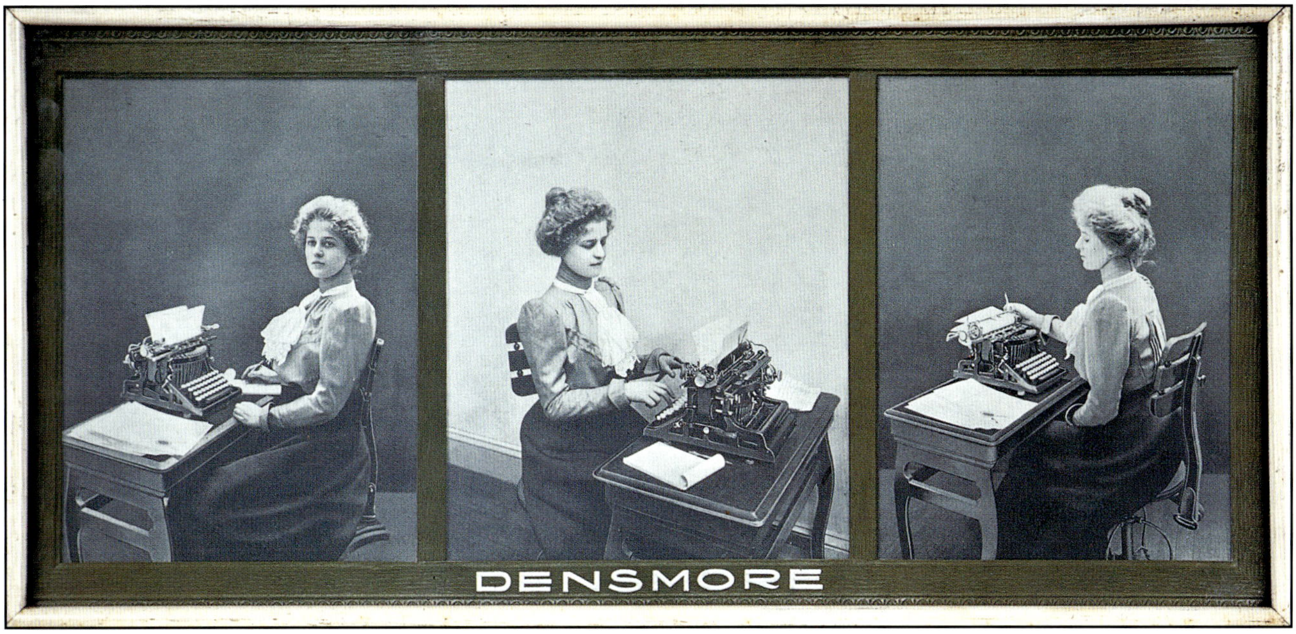

Early Densmore typist, name unknown.

Early Remington typist, name unknown.

Remington Speed typist, Miss Hortense Stolenitz.

Miss Hortense Stolenitz, Remington speed typist (another view).

Early Duplex Typewriter Company typist (c.1917)

Underwood Typewriter Company calendar girl typist (c.1910).

Underwood Typewriter Company calendar girl typist (c.1914).

Remington Typewriter typist from the Central Bureau of Engineering in New York, New York, name unknown.

Remington Typewriter typist: "Miss Remington." Name unknown.

Remington Typewriter typist: "Comrades" (c.1903). Name unknown.

Monarch Typewriter Company typist. Name unknown.

Remington Typewriter typist: "Miss Remington." Name unknown.

Smith Premier Typewriter Company typist. Name unknown.

Royal Typewriter Company typist. Name unknown.

Elliott-Fisher Typewriter Company typist. Name unknown.

Virotyper typist. Name unknown.

International Typewriter Company typist. Name unknown.

Japanese typist.
Name unknown.

Japanese typist. Name unknown [another view].

Royal Typewriter Company typist at the typesetting machine. Name unknown.

Dionne Quintuplets and their Remington Portable Typewriters. Each typewriter had the owner's name shown.

Betty Grable (movie star) at the Remington Model 17 typewriter.

Betty Grable (movie star) at the Remington typewriter advertising her movie "Miss Pilgrim."

Shirley Temple (young movie star) at an Underwood typewriter.

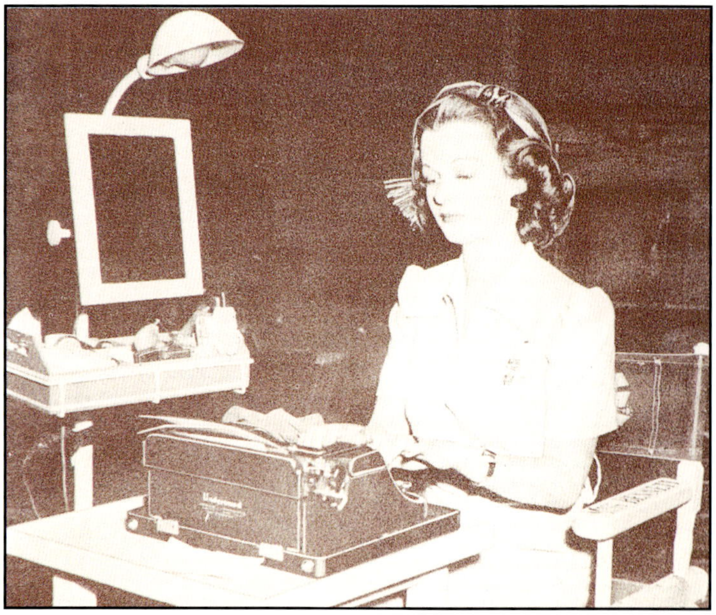

Joan Bennett (movie star) at an Underwood typewriter.

Jane Withers (movie star) at an Underwood typewriter.

Kate Smith (singer) at an Underwood typewriter.

Olivia DeHaviland (movie star) at an Underwood typewriter.

Mary Martin (movie star and singer) at an Underwood typewriter.

Roma Paulsen, World's Amateur Champion, able to type five (5) stroke words a minute for one (1) hour on an Underwood (equal to 125 wpm).

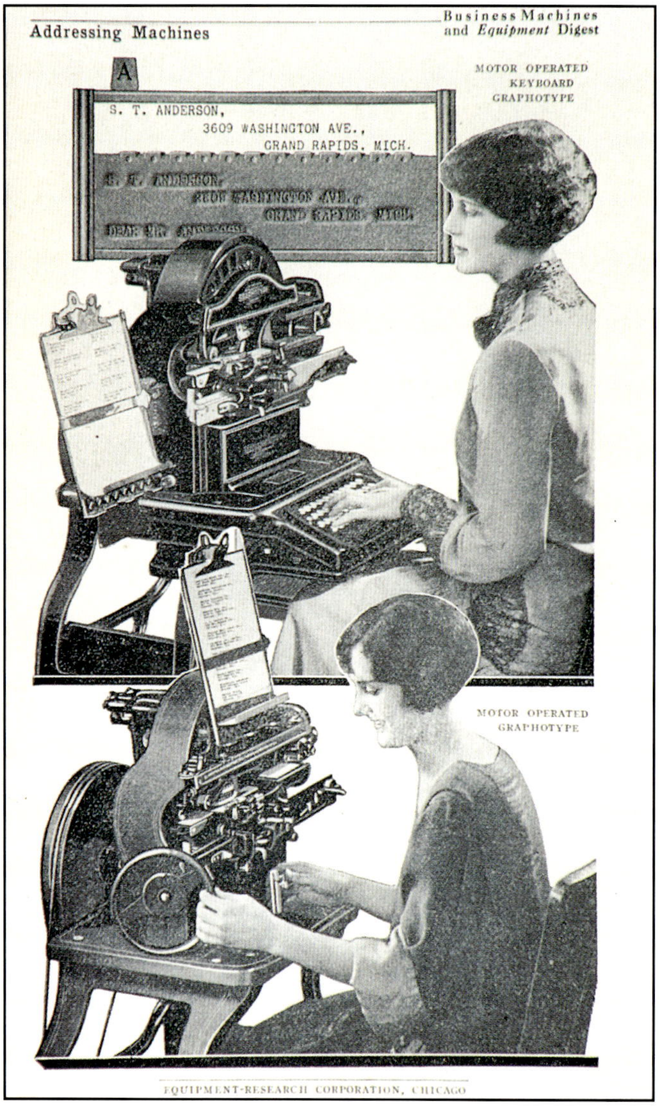

A. B. Dick typist at an addressing machine.

Salome L. Tarr secretary to President Woodrow Wilson at her Remington Typewriter. (c.1913)

E. Remington & Sons typist at their exhibition booth at the 1876 Philadelphia, Pennsylvania Centennial Celebration.

MISS MARGARET B. OWEN
For the Second Time World's Champion Typist
Record on an UNDERWOOD
136 net words a minute for one hour

Miss Margaret B. Owen, world champion typist, on an Underwood sustaining 136 net words per minute for one hour.

Miss Marion C. Waner, amateur typing champion (c.1922).

MISS MARION C. WANER
WORLD'S AMATEUR TYPEWRITING CHAMPION, 1921

Redactron Dual Magnetic Card Word Processing Machine operator and Customer Service Representative (CSR), Miss Sonya Zawalski. (c.1976) Miss Zawalski taught word processing skills to secretaries and typists on a variety of applications.

Miss Remington from *Remington Notes* of 1915.

Tolstoy's daughter and typist.

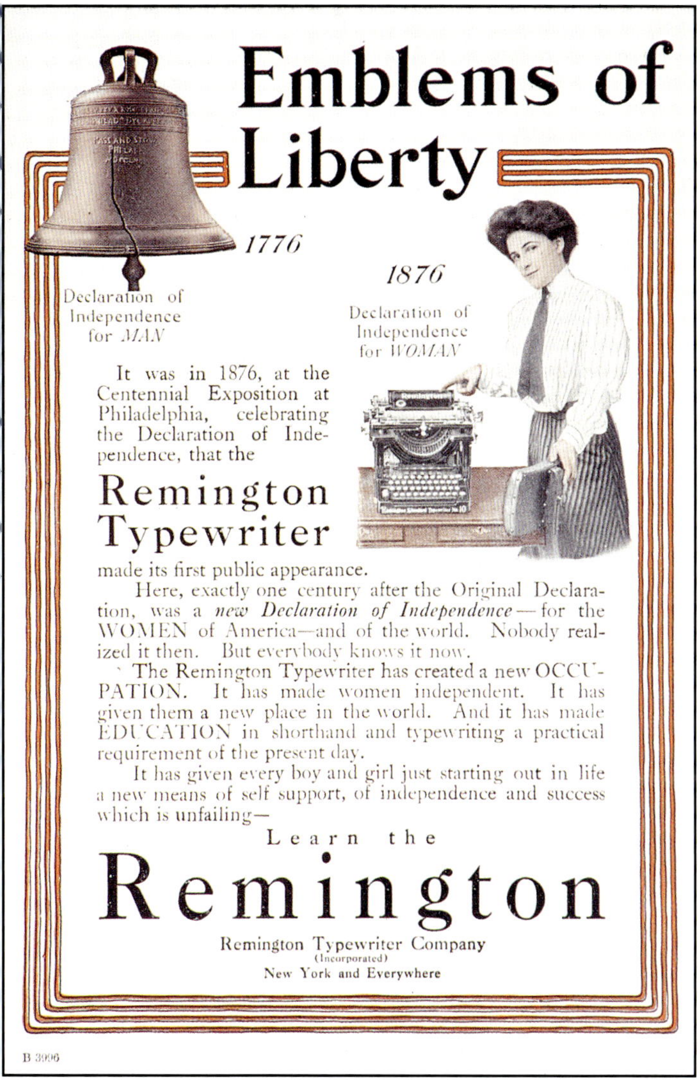

Advertisement commemorating the Women's Declaration of Independence. Notice the Liberty Bell depicting the independence of men and the 1876 Typewriter depicting the independence of women.

Remington Rand typist (c.1981) Last advertisement by Remington Rand. Appeared in the *Wall Street Journal* dated December 18, 1981 . Original copy supplied by Anthony Casillo.

BIBLIOGRAPHY

Anthony, Susan Brownell. *Dictionary of American Biography.* American Council of Learned Societies. Edited by Allen Johnson. Volume I, pgs. 318 – 321. Charles Scribner's Sons, New York: 1928.

Beach, Alfred Ely. *Dictionary of American Biography,* Volume II. American Council of Learned Societies, edited by Allen Johnson. New York: Charles Scribner's Sons, New York: 1929.

Burt, William Austin. *Dictionary of American Biography,* Volume III American Council of Learned Societies, edited by Allen Johnson. New York: Charles Scribner's Sons, New York: 1929.

Durant, Will. *Our Oriental Heritage.* Norwalk, Connecticut: The Easton Press, Norwalk, Connecticut (by permission of Simon & Schuster, Inc.), 1935, 1992.

Durant, Will. *The Reformatio, A History of European Civilization from Wyclif to Calvin: 1300-1564.* Norwalk, Connecticut: The Easton Press, Norwalk, Connecticut (by permission of Simon & Schuster, Inc.), 1957, 1992.

Francis, Samuel Ward. *Dictionary of American,* Volume VI. American Council of Learned Societies, edited by Allen Johnson. New York: Charles Scribner's Sons, New York: 1931.

Gutenberg, Johannes. *The Columbia Encyclopedia.* Sixth Edition. 2001. Bartleby.com

Hatch, Alden. *Remington Arms – In American History.* Toronto, Canada: Clarke, Irwin & Company, Ltd., 1956.

Herkimer County Historical Society . *The Story of the Typewriter, 1873-1923.* Herkimer, New York: 1923.

Hill, Henry Chase. *The Wonder Book of Knowledge.* Philadelphia, Pennsylvania: The John C. Winston Company, 1917.

Knight, Edward H. *American Mechanical Dictionary.* Volume III. Boston: Houghton, Mifflin and Company. Cambridge: The Riverside Press, 1881

Lee, Clark & Henschel, Richard. *Douglas MacArthur* . New York: Henry Holt and Company, 1952.

INDEX